Shibeshi Gebeyehu Belay

Cadastral Procedure and Spatial Framwork for Amhara Region, Ethiopia

Shibeshi Gebeyehu Belay

Cadastral Procedure and Spatial Framwork for Amhara Region, Ethiopia

Südwestdeutscher Verlag für Hochschulschriften

Imprint
Any brand names and product names mentioned in this book are subject to trademark, brand or patent protection and are trademarks or registered trademarks of their respective holders. The use of brand names, product names, common names, trade names, product descriptions etc. even without a particular marking in this work is in no way to be construed to mean that such names may be regarded as unrestricted in respect of trademark and brand protection legislation and could thus be used by anyone.

Cover image: www.ingimage.com

Publisher:
Südwestdeutscher Verlag für Hochschulschriften
is a trademark of
International Book Market Service Ltd., member of OmniScriptum Publishing Group
17 Meldrum Street, Beau Bassin 71504, Mauritius

Printed at: see last page
ISBN: 978-3-8381-5112-0

Zugl. / Approved by: University of Natural Resources and Life Sciences, Vienna (BOKU), 2014

Copyright © Shibeshi Gebeyehu Belay
Copyright © 2015 International Book Market Service Ltd., member of OmniScriptum Publishing Group
All rights reserved. Beau Bassin 2015

Acknowledgements

I gratefully acknowledge the financial support I received from the OEAD to pursue the PhD study. The scholarship includes field studies and conference grants. The work would have been impossible without the generous financial support from OEAD. I have to thank my longtime friend Dr. Zerfu Hailu for creating the contact with OEAD and BOKU and later for his active involvement to solve many problems encountered to realize the scholarship. Without his encouragements I would not be dare to start this work.

My sincere thanks go to the Department of Landscape, Spatial and Infrastructure Sciences, Institute of Surveying, Remote Sensing and Land Information of BOKU for giving me a chance to make the study and for continuous support and encouragement from all staff. I have to send special thanks to my supervisors, Helmut Fuchs (Ao.Univ.Prof. Dipl.-Ing. Dr.techn) and Reinfried Mansberger (Ass.Prof. Dipl.-Ing. Dr.nat.techn), for their rigorous corrections on drafts and guidance during the whole study period. In general, this work would not be in the current shape without the dedication and friendly support from my supervisors.

The entire work deals with developing suitable methods for the second level certification program in ANRS of Ethiopia. Bureau of Environmental Protection, Land Administration and Use (BoEPLAU) is the main responsible institution for the task. I have to thank the office in general and the regional staff in particular for believing in my proposals and for their unlimited moral and material support. I have to specially thank my friends and colleagues at BoEPLAU, Ato Bayeh Tiruneh (the bureau head), Ato Sintayehu Derese (the process owner), Ato Tesfaye Ashinie, Ato Getahun Alameneh, Yegeremew Alemu and many more experts (difficult to list them all) for their active engagement and support during the field work. I have also to thank Mr. Tomas Dubious, the technical advisor, with whom I shared office facilities during my field work. I got the opportunity to discuss and benefit from his skills and experience both in office and in the field. He supported me with very important inputs for the study. Very special thanks to Habtamu Seneshaw (surveyor) for taking the risk to avail the survey equipment for the fieldwork without any complain. BoEPLAU supplied all the drivers and transportation facilities and I sincerely thank both, the institution and the staff, for their support.

I have to thank staff members from the zonal departments and Woreda (district) offices of the study area for taking part in the expert panel discussions and questioner surveys. I have also to thank the staff for their facilitation work during fieldwork and giving me the opportunity to access all necessary official documents and data. Many thanks to the representatives of the major stakeholder offices at Amhara national regional state level for giving me opportunity to discuss with them and for giving me the information and feedback about the interests of the respective institutions to be used as input for my work

My deepest gratitude goes to the landholders (men and women) and land administration committee leaders in ANRS (difficult to name all). Their trust and support to me started since the introduction of the land administration system as a pilot and continued up to now. I have to confirm that it is a pleasure to serve rural landholders in ANRS of Ethiopia. I have to especially thank the selfless land administration committee leaders, such as Ato Azene Zelek (the first elect land administration committee chairperson). They sacrificed their most valuable assets, time and skill, for the realization of the system since then. Without their contribution there would have been nothing to discuss and study now. I have to express my deepest pleasure to get the opportunity to work with the well-seasoned surveyor and international expert Mr. Lennart Backstrom. His contributions and encouragements were priceless.

I have to specifically to thank respondent farmers for individual interviews, farmers who actively attended the group discussion, the key informants, and in general all who contributed for the study. Last but not least I have to thank my family for their endurance and ultimate support during the whole study period.

Preface

The design and the implementation of a progressive land administration system (LAS) in the developing world is a challenge and may be lifetime experience for some people. Land administration system is the expression and management of the relationship between land and human kind. The relation is a must as long as society exists. The relation can be formal or it may be informal and customary. The relationship is affected by socio economic development of a society and influenced by the cultural and political events, within and around the given society. The objective of land administration systems is different for different locations and societies and even across time in the same society. This show on one hand the complexity of the systems and on the other hand how localized and dependent on site specific situations.

It is hard to imagine the best system that can serve the whole world or even a region or even particular areas such as sub-Sahara Africa. Problems related to land are very much of localized nature and need site-specific solutions. This magnifies the need for the development of a progressive system for a given area with distinct legal, political, cultural and socio economic setting. It is hard or may be impossible to compare systems and come up to a conclusion that one is better than the other. However, this does not mean that there are no failed or successful systems in the world. The successful systems are able to manage land to humankind relation in a proper way for a society. The failed systems cannot deliver the particular objectives for a society. Because land is the key ingredient in the national economy, the failure and success of land administration systems is also manifested in the development stage of societies. Successful development of a country and an efficient land administration system has direct and strong correlation.

One can challenge the importance of developing international frameworks, if land administration systems are so distinct and isolated. Nevertheless, regardless of the need for site specific solutions there are many conceptual frameworks from which each system can learn and be adapted to a particular situation. The exchange of knowledge can be done at the conceptual level. For that to happen, proper understanding and explanation of systems is necessary. Land administration systems have to be understood properly and have to be considered as a system. The evaluation of progressive land administration systems is important to understand the contribution of a system for sustainable development.

Models representing the real world are necessary to explain complex issues, such as land management. Sustainable development is an emergent outcome for land administration systems. The Amhara National Regional State (ANRS) formal and informal property right systems are compared using developed and customized legal cadastral domain models. The result shows more similarity than differences between the formal and informal property right systems in the ANRS. The formal system in ANRS has very strong public support. One of the major reasons to get such strong public support is believed to be the similarities of the formal and informal system. The land administration system of ANRS was evaluated to learn from both strengths and weaknesses.

An evaluation framework tailored to progressive land administration systems in the developing world that is elaborated and applied for the ANRS land administration, was designed. Methods were developed by research, training, adoptive implementation and proper feedback. The evaluation framework includes the evaluation of the status of policy and law aspects, the effectiveness of involved institutions, the implementation status of core land-administration functions, the influence of external factors, and the status of inbuilt monitoring and evaluation mechanisms.

The objective of the rural land administration system in ANRS is tenure security contributing for sustainable development. Recognition of the landholding rights of small scale farmers will enhance tenure security and long term investments. The region issues primary books of holdings to 98% of the land holders and planned to start large scale mapping campaign to issue second level certificates with cadastral maps connected to the national grid.

Cadastral surveys, mapping and issuance of second level books of holdings have been identified as key functions of ANRS rural land administration system for the near future. The key functions, which have to be implemented properly, require the enacting of a new cadastral and registration proclamation, the densification of geodetic control points and the selection of an efficient methodology for cadastral surveying. The study aimed at deeper investigation of ANRS rural land administration system by describing the system using legal cadastral domain model and by conducting systematic evaluation. Based on better understanding on the systems requirements the key contributing factors namely; institutional set up, cadastral and registration proclamation, geodetic control points and cadastral survey methods for second level certification program were investigated in the current thesis.

A tool to guide the development of cadastral and registration proclamation is developed. Important provisions identified in the study are:

1. General provisions;
2. Provisions to consider private interests on land;
3. Provisions on management interests of the state;
4. Provisions on access interests of the state;
5. Transitional and concluding provisions.

Details of each of the major provisions have been addressed in this part of the study.

The existing two main geodetic networks, Amhara network (AM) and Ethiopian Mapping Agency (EMA) network, were checked for accuracy, accessibility and status, and point description. The study also identified Precise Point Positioning Systems (PPP) as cost effective method for densification of geodetic control points to be applied for second level certification program in ANRS.

Investigations of cadastral surveying methods for progressive land administration systems had been carried out in the current thesis. The study also considered the major requirements, which have to be fulfilled to launch and roll out cadastral projects in appropriate time.

The research is aiming to improve ANRS rural land administration system. Lessons from practices in ANRS rural land administration system can be adopted to the other four regional states in Ethiopia that are under similar situation. The proposed methodologies to address each key function of LAS are presented as a separate chapter in this study. The methodology, the tools and conceptual framework can be used for newly developed land administration systems around sub-Sahara Africa.

I hope this research is useful indeed in many places where progressive land administration systems are needed and its improvement or implementation is under discussion.

Thesis Structure

This doctoral thesis has five sections. The first section includes the introduction and literature review. The introduction highlights the contribution of a land administration system to sustainable development and identifies the core functions of a land administration system.

Status and the future demands on land administration systems are also presented in the introduction. The literature review gives deeper understanding to evaluation methods for progressive land administration systems, to formal and informal property right systems, to tools for the development of cadastral and registration proclamation, to geodetic control points, and to surveying methods for implementing a rural cadaster. The second section outlines the objectives and defines the research topics. The third section, the methodology part, explains the research area, the research design, the data collection, and the analysis of data. Results and discussions are presented in section four. Section five draws conclusions and implications of the study. References, curriculum vitae and appendix are added at the end.

Abstract

As part of the country's five years growth and transformation plan, currently Ethiopia is implementing a land administration system (LAS) carried out in two development stages. In the first step of the certification program, the legal relation between parcels and their landholders was registered. The second step – the mapping of parcels – will be launched in the near future.

The objective of this thesis is the development of suitable methodologies for the second level certification program in the Amhara National Regional State (ANRS) of Ethiopia. This requires a reality check of the existing situation. Within the study a toolbox was developed covering the institutional set up, the cadastral and registration proclamation preparation, the densification of national grid points, and proper land surveying methods in ANRS.

The core legal cadastral domain model was used to describe both formal and informal settings in a land administration system. The CLCDM was adapted to the situation in ANRS.

In Ethiopia large scale cadastral projects are planned country-wide. As cadastral and registration proclamation is not enacted to facilitate and guide the implementation of cadastral projects, a tool was developed which can be used for the development of cadastral and registration proclamation for rural land administration in ANRS.

Cost effective remote sensing and ground surveying techniques were investigated for their feasibility to produce cadastral maps of different holding types satisfying the needs of users and being connected to the national grid. The study identifies that trust on a system, dependability, and traceability is more important than geometric accuracy.

The study deals with the development of methods mainly suitable for ANRS. But the results and findings of the current thesis largely can be used for the development of a LAS in other regional states of Ethiopia and even in other states of the developing world.

Key words: Land administration system, Sustainable development, cadastral maps, grid control points, ANRS, Ethiopia

Zusammenfassung

In Äthiopien wird derzeit im Rahmen des fünfjährigen Wachstums- und Umsetzungsplans ein Landadministrations-System in zwei Entwicklungsstufen implementiert. In einem ersten Schritt des Zertifizierungsprogramms wurde die rechtliche Beziehung zwischen Grundstücken und Besitzern registriert. Der zweite Schritt – die Kartierung von Parzellen – wird in nächster Zukunft gestartet werden.

Das Ziel dieser Doktorarbeit war die Entwicklung von geeigneten Methoden für die zweite Stufe des Zertifizierungsprogramms in Amhara, einer Verwaltungsregion in Äthiopien. Dies erforderte auch eine Prüfung der derzeitigen Situation. In der Arbeit wurden Werkzeuge entwickelt, welche die institutionelle Umsetzung, die Aufbereitung der rechtlichen Grundlagen für Kataster und Grundbuch, die die Verdichtung des Festpunktfeldes und die für die Kartierung geeigneten Vermessungsmethoden umfassen.

Das Kernmodell eines rechtlichen Katasters wird zur Beschreibung der formalen und informalen Gegebenheiten in einem Landadministrationssystem verwendet. Dieses Modell wurde für die spezifische Situation in der Amhara-Region adaptiert.

Kosten-effiziente Fernerkundungsmethoden und terrestrische Vermessungsverfahren wurden auf ihre Eignung zur Produktion von Katastralmappen von unterschiedlichen Besitzverhältnissen untersucht. Diese sollen die Anforderungen der Nutzer erfüllen und an das nationale Koordinatensystem gekoppelt sein. Die Studie zeigt auf, dass Vertrauen, Zuverlässigkeit und Nachvollziehbarkeit für ein Katastersystem wichtiger sind als die geometrische Genauigkeit.

Die Studie behandelt vorrangig die Entwicklung von Methoden für die Amhara Region. Aber die Ergebnisse und Erkenntnisse dieser Studie können zum Großteil auch für die Entwicklung von Landadministrationssystemen in anderen Regionen Äthiopiens und in den Staaten des Südens verwendet werden.

Schlüsselwörter: Landadministrationsystem, Nachhaltige Entwicklung, Katasterkarten, Festpunkte, Amhara Region, Äthiopien.

Table of contents

1	**Introduction**	9
2	**Definition and Objective of the Research Topic**	13
2.1	Definition of the Research Topic, Hypothesis and Research Questions	13
2.2	Objectives of the Research	14
3	**Literature Review**	15
3.1	Evaluation Methods for Progressive Land Administration Systems	15
3.2	Formal and Informal Property Right Systems	18
3.2.1	History of the Formal System	18
3.2.2	History of the Informal System	19
3.2.3	Legal Cadastral Domain Model	20
3.3	Toolbox for Managing Interests on Land	21
3.4	Cadastral and Registration Law	22
3.5	Reference Points for Cadastral Surveying	23
3.6	Selection of Cadastral Surveying Methods in Progressive Land Administration Systems	25
3.6.1	Review of experiences and practices	25
3.6.2	Selection of Cadastral Methods	26
4	**Study Area, and Methods**	28
4.1	Study Area	28
4.2	Methods	28
4.2.1	Individual Interviews	29
4.2.2	Questionnaire Survey and Expert Panels	30
4.2.3	Discussion with Major Stakeholders	30
4.2.4	Identification and Evaluation and Static Measurements of Reference Points	30
4.2.5	Comparisons of Survey Methods Using Sample Measurements	32
4.2.6	Compiling and Processing of Findings	33
5	**Results and Discussion**	34
5.1	Evaluation Result of Land Administration System in ANRS	34
5.1.1	Policy and Law	35
5.1.2	Institution and Management	39
5.1.3	Operational Level	42
5.1.3.1	Land Tenure	42
5.1.3.2	Land Valuation and Expropriation	46
5.1.3.3	Land Use Planning	48
5.1.3.4	Development Control	48

5.1.4	External Factors	48
5.1.5	Review Processes	51
5.2	Description and Comparison of the Formal and Informal Property Right Systems	58
5.2.1	The Hierarchy of the Legal System in Ethiopia	58
5.2.2	Formal Landholding Right	59
5.2.3	Informal landholding Right	64
5.3	Points to be Considered during Preparation of Cadastral and Registration Proclamation for Second Level Certification Program in Ethiopia	69
5.3.1	Users' needs and interests on land	69
5.3.2	General Provisions	70
5.3.3	Provisions to Consider Private Interests on Land	71
5.3.3.1	Provisions to Describe the Rights of the Landholders	71
5.3.3.2	Provisions to Describe the Responsibilities of the Land Holder	72
5.3.3.3	Provisions to Define the Restrictions on the Land Holder	72
5.3.4	Provisions on Management Interests of the State	73
5.3.5	Provisions on Access Interests of the State	73
5.3.6	Transitional and Concluding Provisions	74
5.4	Geodetic Control Points	74
5.4.1	Network Established by BoEPLAU (AM network)	75
5.4.1.1	Accessibility and Status (AM network)	75
5.4.1.2	Point Description (AM network)	76
5.4.1.3	Accuracy (AM network)	77
5.4.2	Network Established by Ethiopian Mapping Agency (EMA Network)	78
5.4.2.1	Accessibility and Status (EMA network)	79
5.4.2.2	Point Description (EMA network)	82
5.4.2.3	Accuracy (EMA Network)	85
5.4.2.4	Long Duration GNSS-measurements	87
5.4.3	Recommended Method for the Densification of the Ground Control Point Network	89
5.5	Selection of Cadastral Survey Methods	90
5.5.1	Necessary Considerations before the Implementation of Cadastral Survey Projects	90
5.5.1.1	Objective and Goal Setting Decision Criteria	90
5.5.1.2	Timing and need assessment criteria	92
5.5.1.3	Way of working design criteria	93
5.5.2	Selection of Proper Surveying and Mapping Techniques	95

6	**Summary, Conclusions, and Recommendations**	**101**
6.1	Identification of the Key Intervention Areas for Effective Development of Progressive Land Administration Systems Using ANRS Rural Land Administration as a Case	101
6.1.1	Evaluation Progressive Land Administration Systems	101
6.1.2	Formal and Informal Property Right Systems	102
6.2	Toolbox Guiding the Development of Methods for Identified Key Intervention Areas of Progressive Land Administration System	102
6.2.1	Identification of Tools for Organizing Effective Institutional Set up	102
6.2.2	Tools for the Development of Cadastral and Registration Proclamation	102
6.2.3	Tools for Geodetic Control Points for Cadastral Surveying	103
6.2.4	Tools for Selection of Cadastral Surveying Methods	104
7	**References**	**105**
8	**Index of tables**	**112**
9	**Table of figures**	**114**
10	**Appendix**	**115**
10.1	Discussion Points with the Land Administration Committee Members	115
10.2	Discussion Points with Selected Woredas, Zones and Regional Land Administration Professionals	117
10.3	Questioner for Individual Farmers	119
10.4	Questioner to be Filled by Major Stakeholders of Land Administration System in ANRS of Ethiopia	121
10.5	Questioner to be Filled by Land Administration Offices	122
10.6	Accuracy Comparison of AM Point Coordinates and PPP Measurements	133
10.7	Root Mean Square Error for Different Observation times [in ± m]	135
10.8	Velocity Factor of AM Points	137
11	**Table of abbreviations**	**139**
12	**Curriculum Vitae**	**141**

1 Introduction

Land is one of the most important assets for sustainable rural development all over the world (Burns, 2007; De Soto, 2000). Land and human power are identified as the two available resources for development in Ethiopia (MoFED, 2010). Land administration system is the key input for proper management of humankind to land relations. The relation of humankind to land is well managed in the developed world. On the contrary, the coverage and status of formal land administration systems in the developing world is minimal.

The rules to govern the management of this resource are determinant for societal development. The development of a society is a dynamic and continuous process that has resilient impact on the nature of the relationship between human race and its land. The relationship between people and land can be spiritual or metaphysical and material (Sheehan, 2001), partially documented by a land tenure or land administration system.

Cognizant to the fact that a land administration system is the key input for sustainable rural development, many donor driven large scale land administration projects were introduced in the developing world in general and in Africa in particular. Unfortunately the outcomes of these projects were not satisfactory. The reasons for failure are numerous, different, and manifold. But the common factor is that they are all imported and attempted to introduce standard land administration systems from the north to south.

The emergence of a traditional property right system in Ethiopia, especially in the ANRS, is unswervingly related to the management of scarce rural land and other natural resources. The introduction and wide scale application of sedentary agriculture was the prime cause for a gradual decline of shifting cultivation in the ANRS. The Irist system – a traditional tenure based on blood related group of people – was the dominant property right system all over the ANRS up until the end of the Imperial regime in 1974 (Adal, 2002; Ashenafi & Leader-Williams, 2005; Rahmato, 2005). The military junta – called Derg – abolished the Irist system and introduced public ownership of land. The current government, after taking power in 1991, enacted a new land administration and land use proclamation. Land redistribution was banned by law. But land in Ethiopia continued to be a public property. The aim of the land policy of the present day in Ethiopia is attaining tenure security for sustainable rural development.

The emerging progressive land administration system in ANRS is the focus of this thesis. Progressive land administration system in the context of this thesis is used to refer to a land administration system capability to address dynamism in economic, social, environmental, and political circumstances, and to contribute to sustainable development of a given society. The progressive land administration system deal with core functions of land administration system, namely tenure value, use and development (Williamson, et al., 2010).

Activities in progressive land administration systems will be prioritized and implemented on step by step basis. The progressive land administration system in the ANRS has primary and second certification levels. The strategic plan with two major parts was developed based on lessons learnt from pilot projects. The first stage dealt with adjudication, first registration and issuance of primary book of holdings. The second stage of the strategic plan will focus on the development of a geo-referenced spatial description for all holdings step by step.

The book of holdings will be upgraded from primary to second level by attaching geo-referenced parcel maps to the land book. Computerization of the analogue land register is a pre-requisite for upgrading the book of holdings to the second level. The software for the computerization of the land register in the ANRS is called Information System for Land Administration (ISLA) and was developed by BoEPLAU (Bureau of Environmental Protection Land administration and Use) with support from Swedish International Developing Agency (SIDA). In the past ten years the ANRS managed to register and to certify more than 3.6 million properties in the first level. The second level of certification program is about to start.

Ethiopia developed a five years growth and transformation plan (GTP), targeting on attaining sustainable rural development (MoFED, 2010). One of the targets in the five years growth and transformation plan is the establishment of an efficient land administration system and the issuance of second level book of holdings for all small scale farmers in four populous national regional states of the country (Amhara, Tigray, Oromea, and Southern regions). The target is directly related to the mapping of parcels and to the registration of different interests on land.

The objective of the thesis is to develop a toolbox suitable for the development of methods for the second level certification program. The development is based on systematic evaluation results and on an adapted description of the legal cadastral domain model (Paasch, 2012). The toolbox approach is chosen due to its capability to be flexible during implementation and due to its potential to consider different scenarios.

The endeavor of developing toolbox suitable methods (tools) for second level certification program in ANRS of Ethiopia has two interlinked objectives: The identification of key intervention areas for a progressive land administration system and the preparation of a toolbox that can guide the development of suitable methods for the identified key intervention areas.

Critical evaluation of strengths and weaknesses of the system is necessary before launching the large scale second level certification program. But a suitable evaluation framework to address also the site specific nature of progressive land administration systems was lacking (Steudler et al., 2004). Therefore, an evaluation framework for progressive land administration systems is developed and used to evaluate ANRS rural land administration system.

Strengths and weakness of the ANRS land administration system are evaluated. Results of the evaluation pointed out the key elements to be investigated. A toolbox to guide the development for the implementation of the identified key elements is developed. The identified key elements for second level certification program in Ethiopia are: the enacting of cadastral and registration proclamation, the establishment and densification of ground control points, the selection of suitable surveying and cadastral methods, and the institutional set up.

The systematic description of the ANRS land administration system is conducted using the legal cadastral domain model (LCDM). The model was adapted based on the information from the formal land administration system in ANRS. The new equivalent model for the informal setting is developed based on the experience from the ANRS. The formal and informal settings are described and compared for proper understanding of the system. Proper understanding of the existing system is a must for both, the development of applicable tools that can be easily practiced during the planned second level certification program and for sharing developed information, experience and knowledge between systems.

Cadastral projects are different from other technical surveys as they deal with legal property boundaries. The holding rights in the case of ANRS were adjudicated during first level of the certification program. The book of holding is the sole legal proof for the holding right. The interests on land and the management of these interests using cadastral maps need to be clearly defined by cadastral and registration proclamation. In this thesis a tool is developed to guide the preparation of cadastral and registration proclamation that is one of the requirements for the planned second level certification program in Ethiopia. The tool deal with the legal boundary types, with the accuracy required for different holding types and with the need to connect the cadastral mappings to the national grid.

Connection to the national grid is necessary for a multipurpose cadaster. The cadastral maps can serve for planning and construction of infrastructure and services, if they are connected to the national grid. In this way road networks, power and telecom lines, water supply and drainage etc. can be effectively planned and managed. The connection to the national grid is carried out by the network of geodetic ground control points. The availability and the distribution of ground control points have impact on the costs and on the speed of cadastral mapping.

Ethiopian mapping agency (EMA) is the legal authority for the countrywide creation and maintenance of geodetic control points (EoE, 1980). Due to urgent need of geodetic information for the establishment of modern land administration system in the ANRS, the Bureau of Environmental Protection Land Administration and Use (BoEPLAU) took initiative and created with technical and financial support from SIDA additional geodetic reference points (Miskas & Molnar, 2010).

The monuments of geodetic control points created by EMA are very old, difficult to locate and not properly maintained. Point descriptions are also very poor and very obsolete to the extent that they can no more be used to guide a user to reach them. The accuracy is also problematic, as some of them have been moved, destroyed or significantly damaged to an extent that they cannot serve their purpose.

In addition to sparse distribution, the geodetic points are located mostly on inaccessible hill tops and they are not reliable as they are rarely attended since many years. As part of this study, BoEPLAU points (AM network) and EMA points (EMA network) were controlled and evaluated considering the availability, the point description, and the accuracy level. Recommendations are elaborated, how the establishment and surveying of additional ground control points could be realized in future.

The coverage of well-functioning cadastral systems in the world is still confined to the developed world (Williamson & Ting, 2001). The status of country's economic development and the availability of well-functioning cadastral systems seem strongly correlated (Steudler, et al., 2012). Unfortunately, in the last decades, no significant improvement was recorded on the cadastral systems of the developing world. Their suitability for sustainable economic development in the light of today's technology and the present social and economic requirements of each country cannot be proved. Especially in the developing world, the main criteria to be considered to choose from cadastral surveying techniques during the implementation are cost, time, accuracy, and ease.

Before confronting challenges emerging from complications of untimely and copied establishment of spatial infrastructure, it is detrimental to look for options and carefully choose location specific solutions using well-defined selection criteria. Except some trial and piloting reports' dealing with specific outputs, there are no established selection criteria for the establishment of spatial infrastructure in progressive land administration systems. The commonly applied cadastral surveying techniques in Ethiopia can be categorized into two: Remote sensing tools and ground survey tools. As part of this thesis cadastral methods suitable for second level certification program were investigated and recommendations for a method to be applied in ANRS are given.

2 Definition and Objective of the Research Topic

Currently the second level certification program in ANRS, including cadastral mapping, is at the planning stage. There is a lot of work to issue second level books of holdings for small scale farmers and map all parcels in the whole area of ANRS. Before launching such a huge and demanding task it is natural to evaluate the status and identify the key intervention areas to reach to the goal.

Systematic evaluation of the existing ANRS rural land administration system was not conducted so far. Proper tools to evaluate the system were lacking

BoEPLAU is the responsible institution for the implementation of rural land administration system in ANRS. It is responsible for environmental and land administration activities. The institution is all in one type with legal mandate for both cadastral surveys and land registration related activities. However, the institutional set up is not appropriate to manage core land administration functions.

Geodetic control points are necessary to connect cadastral maps with national and international grids. Ethiopian mapping agency is the responsible organization for creation and maintenance of geodetic points in Ethiopia. The density and distribution of these points is not satisfying such a large scale project.

Cadastral surveys in ANRS were limited to pilot project sites and large scale irrigation projects. Different ground survey tools ranging from hand held GPS code measurements to high precision total stations and RTK GPS were used in different places. Remote sensing tools, such as orthophotos and high resolution satellite imageries were also tried in different pilot projects. Until yet, there is no consensus about the method to be applied for the cadastral mapping.

2.1 Definition of the Research Topic, Hypothesis and Research Questions

The hypothesis is that – in spite of the different social, political, and administrative background of each country – it is possible to develop a toolbox to guide the development of methods for progressive land administration systems in the developing world, which take economic, social, and environmental dynamism into consideration. The key areas of intervention can be identified by a systematic evaluation of land administration system. The description of the land administration system using an adapted of the legal cadastral domain model of Paasch (Paasch, 2012) is necessary for proper understanding and knowledge exchange.

The case of the Amhara national regional state of Ethiopia is used as example and testing ground. The research will inquire the following two major questions which are related to the problem.

- The first main question is: How can the key requirements for the establishment of a new cadastral system in the developing world identified by using ANRS as a case? It was necessary to ask two sub questions to get an answer for the first question:
  o Can systematic evaluation methods customized and applied to identify gaps and to learn from progressive land administration systems using ANRS as a case?
  o Can formal and informal settings in ANRS be described for effective integration and for deriving lessons for toolbox development?
- The second and the follow up main question is: Is it possible to develop a toolbox that can guide the selection of suitable methods for the implementation of progressive land

administration systems? Also the second main question has sub questions related to the identified key intervention areas:

- What is a suitable institutional setup for effective implementation of progressive land administration system using ANRS as a case?
- What is an appropriate reference framework to enhance progressive land administration system in Ethiopia?
- What are the suitable techniques and technologies that can fulfil the needed low cost, high speed and proper positional accuracy for progressive land administration system using ANRS as a case?
- What are the major requirements and considerations for the development of cadastral and registration proclamation to support second level certification in Ethiopia?

2.2 Objectives of the Research

The study has two interlinked objectives that contribute to the overall objective, which is development of suitable methods for progressive land administration systems using ANRS as a case.

The two objectives of the study are:

- Identification of the key intervention areas for effective development of progressive land administration systems using ANRS rural land administration as a case.
- Designing a toolbox that can guide the development of methods for identified key intervention areas of progressive land administration system.

The above two main objectives have interlinked sub-objectives. Sub-objectives for the first objective are:

- Evaluating the rural land administration system in ANRS using customized systematic evaluation framework to get lessons and identify key intervention areas;
- Description of formal and informal settings in ANRS using adopted legal cadastral domain model for effective integration of the two systems.

The sub-objectives of the second main objective are related to key intervention areas. The four sub objectives are:

- Developing methods for effective institutional set up to manage progressive land administration systems using ANRS as a case;
- Developing methods for the drafting of cadastral and registration proclamation in Ethiopia;
- Identification of suitable method for establishment and densification of geodetic control points in ANRS;
- Preparation of a guiding tool for selection of suitable cadastral survey methods using ANRS as a case.

3 Literature Review

There is a bulk of literature dealing with property right systems. However the publications dedicated on progressive land administration systems and site specific problems of ANRS of Ethiopia are still limited. The focus of this section is to discuss the findings that are related to emerging land administration systems. The section particularly deals with the studies and recommendations that can contribute for the development of effective methods for second level certification program in ANRS. The discussion includes evaluation and description of the existing land administration system to get lessons from strengths and weaknesses of the existing system.

The neo-classical economic theory of property rights asserts identified individual property right systems as the ultimate sources of wealth and increased investment (Deininger & Jin, 2006). On the contrary, the Marxists considered land as a public property. They argued that privatization of land can lead to concentration of in particular land and wealth in general into the hands of few capitalists. Privatization according to communistic thinking is unfair allocation of public property to privileged few (Feder, 1999; Fitzpatrick, 2006).

The debate about land to humankind relationship in Africa is characterized by two 'schools of thought'. The first group argues that land policies should be rooted in a theory of social capital (most African traditional land tenures belong here). The other group is convinced that individualized land tenure systems are more effective and desirable (Obeng-Odoom, 2012). In reality the progressive land administration systems in Africa are in the middle of the two extremes. The participatory nature enables them to exploit the social capital and the formal enforcement mechanism is the reflection of the individualized tenure systems.

The early individualized 'modern' land administration systems were introduced in Africa by colonial powers, but most of them were not sustained because local population considered them as threats. However, the effect of colonization on land tenure in Ethiopia is minimal. Several large-scale individualized land administration projects introduced by international aid and funding institutions failed in Africa. Although many African countries have recently adopted highly innovative and pro poor land laws, lack of implementation thwarts their potentially far-reaching impact on productivity, poverty reduction, and governance (Deininger et al., 2008). The land administration system in ANRS is among the recently adopted systems.

The knowledge and information generated by the local society to satisfy the growing needs is the cause for the emergence and development of property systems. The change in the way of life is triggered by the scarcity of the natural resources to meet the growing needs of the society. The precision of the definition of the property rights and the rigor with which they are enforced is closely related with the value of the resources and population density (Mattsson, 2003; Williamson at al., 2010).

3.1 Evaluation Methods for Progressive Land Administration Systems

The reasons for success and failure of land administration systems are manifold, and that the development of a 'land administration theory' on this matter should be at the top of the research agenda (Van der Molen, 2002). Proper evaluation methods and practice have a potential to convert challenges in to an opportunity for change and experiential learning. Many frameworks and methodologies that attempt to evaluate characterize and assess land administration systems in the world were developed (Chambers, 1983; Cusworth & Franks, 1993; Diallo & Thuillier, 2005). But they could not properly address local problems and situations of progressive land administration systems in developing countries (Burns, 2007). On the other hand an evaluation framework for progressive land administration systems is an urgently needed task in developing countries (Lemmen et al., 2009).

Standardized methods or a quality framework for improving, evaluating or comparing land administration around the world is still lacking (Ali, Tuladhar, & Zevenbergen, 2010). The lack of standardized methods is due to the fact that the land administration systems are dependent on the cultural and social values of local societies in which they operate (Steudler et al., 2004; Williamson & Fourie, 1998). Land issue as one of the most valuable resource of all nations is covered by a constitution or policy level documents. Land issue as a constitutional provision is subject to controversies. The advantage is related to creating stable policy ground and the disadvantage is lack of flexibility once land issues are included in the constitution.

Land policy is among the key policy issues in almost every country. The policy level evaluation of this thesis investigates if the system is well defined by objectives, if it responds to the needs of the society, if it is equitable for all, and if the system is economically viable (Steudler et al., 2004). The policy level evaluation is used to identify the need for policy change and the need for additional proclamations for implementing the planned second level certification program.

When an appropriate legal framework and transparent public administration structures are lacking, land administration can only make the best of a bad job (Van der Molen, 2002). The conventional way of property right definition procedure is a top down legal process (Dale & McLaughlin, 1999; De Soto, 2000). Contrary to this conventional way, rights and obligations on land in the ANRS are defined by participatory adjudication process. The advantage of the conventional way is simplicity for implementation and enforcement. It is also suitable to develop a centralized and uniform system. The main disadvantage is that the system is not capable of involving users and it is poorly organized to address the needs of users on how to manage their property.

Defining a property is a key first step in land policy formulation. 'Property' is the description of the legal relationship with a thing. The property rights can be better described as a bundle of rights to get a room for flexibility to decide the items in the bundle based on the specific needs of each country. The rights included in the bundle of rights are different in different jurisdictions.

The difference in the type of rights included in the bundle per se cannot be sited as a source for insecurity. Rights are classified into access, withdrawal, management, exclusion and alienation. The property right holders are also classified into authorized entrants, authorized users, claimants, proprietors and owners (Ostrom, 1998). The major rights that have protection from the formal legal setting are mostly subject for registration. The aim of an efficient and up to date registration system is to describe the right holder on the land (Hodgson, 2004). Registration system of the right on land is one of the core functions of land administration system. Holding right is subject for registration in ANRS.

Land administration is the public sector activity required to support the alienation, development, use, valuation and transfer of land. Land administration cannot be treated in isolation from other activities. The case for good land administration rests on good commercial grounds as well as up on matters of social justice (Dale and McLaughlin, 1999). Formalization will do little good, if it is not backed up by a coherent legal system and authority structure that promises effective enforcement of the rights inherent in, and implied by, the granting of titles (Bromley, 2008). This fact is in contradiction to Hernando De Soto who claims formalization of rights can change dead capital in the developing world to active capital (De Soto, 2000).

Absolute land ownership is hard to imagine in a society because one can affect other members of the society while he is trying to enjoy the ownership right (Mattsson, 2003). We need to be clear when we refer to ownership or property rights. The explanation should include the type of activities and income streams, the authority to define them and whether they are private or common (Van Den Brink R., 2002). The equivalent right in ANRS is a holding right. Holding right is the right to use land in perpetuity, but the holder has no right to sell. The right to sell land is not part of the bundle of rights in the case of ANRS.

Land tenure is a rule invented to regulate social behavior. The rules define how the land rights are exercised and access to land is granted. In short, land tenure defines, who can use what resources for how long, and under what conditions (FAO, 2002). This is maybe why Marx goes further and uses the term 'relation of production' (Aredo, 2003).The rule of the game is either agreed by the society or the state shall enact it in the form of formal law. The state or the community has to be capable of implementing and enforcing their rules to bring about tenure security (Van Den Brink, 2002).

Institutions are the humanly devised constraints that structure human interaction. They are made up of formal constraints (rules, laws, constitutions), informal constraints (norms of behavior, conventions, and self-imposed codes of conduct), and their enforcement characteristics. Together they define the incentive structure of societies and specifically economies (North, 1993). Based on this definition both formal and informal land administration system settings can qualify to be considered as institutions.

Land administration systems in general and institutional aspects in particular are context dependent and therefore they are different in different societies. The transfer of institutional arrangements was a key driver in the colonization process, where institutions appropriate to one (geographic and social) context were transferred into a new context (Smajgl & Larson, 2007).

The contextual nature of institution can be one of the reasons for failures of land administration systems introduced by colonial powers in Africa. Historical background of land administration institutional arrangements are influenced by factors such as whether the system is decentralized, de-concentrated, or centralized. The level of education and training in a country is also an important factor for the nature of formal land administration institutions (Williamson, 2000).

Experience shows that successful land administration systems have all the land administration functions within one government organization. There should be one government department responsible for the land administration infrastructure in a country. There are examples in developing countries where the institutional arrangements are within one government department (Williamson, 2000). The institutional setup in ANRS is similar to the recommended one.

The target of management level evaluation is to find out the efficiency and clarity of organizational arrangement, if the strategies are appropriate to reach and satisfy objectives, if involved institutions have each clearly defined task, and if they cooperate and communicate well with each other, and if the private sector is involved (Steudler et al., 2004). The management level dealing with institutional setup is one of the important inputs for the effective implementation of the core functions.

Land administration systems cannot be understood, built, or reformed unless the core processes are understood (Williamson et al., 2010). The core processes deal with operational functions. The key attributes of land administration system are land tenure, land value, land use and development control. The attributes can be expressed in the form of four functions: juridical, fiscal, regulatory, and information management (Dale & McLaughlin, 1999).

Land tenure is defined as the mode in which rights to land are held. Value is to be understood as all kinds of values which land might have. Land use is all the kinds of use land might have (Van der Molen, 2002). The definitions of land tenure value and land use are also the same in the context of ANRS. The operational issue basically deals with understanding the functions and core processes of land administration system.

There have been various approaches to evaluate and to compare the performance of land administration projects in developing countries. Each land administration project includes processes for evaluating the effectiveness and impact of the implementation. However, many of these focus on the efficiency of land administration processes and the capacity of institutions (Mitchell et al., 2008).

For the evaluation of an administration system as a whole another two additional areas need to be considered ('review process' and 'external factors') (Steudler & Williamson, 2005). The external factors discussion in the study includes human resource development, capacity building, professional association, and technical developments (Steudler et al., 2004). Review processes consider objectives and strategies, performance and reliability of system, and customer satisfaction (Steudler et al., 2004).

3.2 Formal and Informal Property Right Systems

3.2.1 History of the Formal System

The tenure structure during the Imperial period was quite complicated and in parts of the country highly exploitative (Mesfin, 1991; Ashenafi & Leader-Williams, 2005; Rahmato, 2005). During this feudal period there were on the one hand small-scale owner-cultivators and on the other hand large landholders, who in many instances obtained their possessions through political means. Such landholders were members of the nobility and the local gentry. The nobility were absentee landlords while the gentry resided close to their property. The system is said to be exploitative, because nobility and land lords contribute no part in the production process but demand the highest share of the benefit from agriculture. The shares they demand were ranging from one third to two thirds of the product especially in the southern parts of the country. Complete and irreversible defeat of landlordism was the greatest achievement of the military dictatorship Derg (Rahmato, 2005).

The revolutionary change during the Derg time eradicated all the class relations, but also removed the growth attempts of the enterprising farmers. During the Derg era, the focus of the transformation was highly influenced by equity issues and suppressed the efficiency needed for agrarian development (Rahmato, 2005; Ashenafi & Leader-Williams, 2005).

The periodic redistribution of land, the ban on land renting, ban on the hiring of labor, grain requisitioning, forced villagization and cooperativization were the major burdens on the production system. Nowadays, after they have been abandoned, class differentiation within the peasantry became a thing of the past.

The past rural policies were not in a position to encourage the creative skill and productivity of the landholders. To the contrary, the policy forced them to live in the abject poverty. The rural economy has undergone a shift towards micro-agriculture in the last three decades. Because of the wrong policy, the peasant farm was growing smaller, producing less, and increasingly losing its fertility. The average household gained fewer farm assets and was much more vulnerable. Matters have been made worse by high rates of population growth and severe demographic pressure on the land leading to what one might call the 'saturation of rural space' (Ashenafi & Leader-Williams, 2005).

The past three political upheavals in Ethiopia were strongly influenced by the land issue. One of the most popular mottos of the revolution was 'Land for the tiller' (Meret larashu). On 4th of March 1975, the Derg proclaimed the nationalization of all rural land through the rural land proclamation number 31/1975; since then land is under the ownership of the government. The same proclamation is the base of the establishment of Kebeles (parishes), the lowest administrative units in Ethiopia. The Kebele (parish) administrations were responsible for land reallocation and resolving land related conflicts during the Derg era. The current government included most important land policies in the constitution.

The constitution delegated the details of land issues to be proclaimed separately. Proclamation no 89, later on amended by proclamation 456, was the result of the constitutional provision. In proclamation 456/2005, it is proclaimed that land is not subject for sell or any other type of exchange in Ethiopia. The ownership to land is exclusively vested to the state and to the peoples of Ethiopia. It is only the holding right that is given to individual citizens (FDRE, 2005).

By the framework law, power is given to the regional states to enact their own land administration and use the proclamation in accordance with the federal law. The regional laws are supposed to take into account the site specific conditions and to achieve the regional objectives. The proclamation also enables regional states to establish their own institutions pertinent for the implementation of the proclamation (FDRE, 1995).

The ANRS land law (133/2006) (ANRS, 2006) was developed based on the provisions given by proclamation (456/2005) (FDRE, 2005). Many consultations with the major stakeholders were made before enactment. The needs and interests of different stakeholders were at most considered during the drafting process. The development of the legal system subsumes the need for high technology, participation, justice and proper information dissemination, and training of the land holders. As the system is the combination of complex technical and social issues, the efficacy of the system will increase, if it is developed and implemented in a participatory process.

Proclamation 133/2006 (ANRS, 2006) is an improved version of proclamation no 46/2000 (ANRS, 2000). The law was amended based on the experiences gained during the implementation of proclamation no 46/2000. It attempts to resolve the problems encountered during the implementation and it guarantees better rights for landholders. The ultimate objective of the current proclamation is to attain tenure security and to enable sustainable development.

3.2.2 History of the Informal System

Indigenous land tenure systems in Ethiopia were varied and evolved through a complex of processes. The major forms of land rights and land tenure systems operated in Ethiopia were Atsme Irist and Gult (Ambaye, 2013). The key informants in different sample areas also defined Gult as the right to administer an area, mostly assigned for the members of the royal family. The Gult system was a decentralized taxation system, where the Gult right holder has the right to levy and collect tax on behalf of the central government. The tax used to be paid in kind. Usually the Gult holder adds some margins on the proportion for covering his administrative costs. The commonly used proportion was one tenth of every product. The Gult right holders have the right to exempt Gebars (Irist right holders) from taxes. In general, the Gult right holders are responsible for the overall administration of the area, but they have no right or power to give or take land from Gebars (Mesfin, 1991; Ambaye, 2013).

The Gult system was abolished during the Imperial period after the Italian invasion and replaced by a centralized taxation system, which continues up until the end of the Derg period.

Atsme Irist commonly known as Irist is a hereditary ownership system of land tenure. The Irist holder is the kinship group of the first settlers of the area. The individuals in the kinship group have their own private holdings, but they cannot transfer their right to an outsider without the consent of the group leaders. The Irist system was the most dominant system in Amhara National Regional State.

There was a system of allocating unoccupied land for different government services as a salary for the military and for civil servants. This type of land was known as Maderia. Unlike Irist, Maderia land was not hereditary, though the services sometimes were inherited. The land was reallocated, when the service to the government ended. Maderia covers only the interest in the estate, while Gult is the right to tax and administer land, including Maderia land itself. The government also allocated land for church services. Sometimes the government allocated unoccupied land to the church as a Gult area. Then the church was allowed to collect taxes from users as a compensation for running the church services.

The church services givers are divided mainly into two groups: firstly the priests and deacons, locally called Kedash, and secondly the locally called Debteras, who are assigned and educated to render the administrative and educational services.

In almost all cases the land allocated for the church services is within the vicinity of the church so that the service giver is easily available whenever required. Most of these lands are still in the hands of the churches, even if the management system now is quite different. All the land allocated for church services are tax exempt.

The individual land tenure situation before the land reform proclamation may be broadly described as follows – apart from different types of land allocated for different types of social services:

- People had no right to own land (such as an artisan groups);
- Landless peasants were operating as tenants;
- Gebar or Irist holders operated their own small plots of land;
- Landlords, mostly absentee landlords, had some thousands of hectares; and
- Commercial farmers – originally not from the farming community – were operating modern large scale farms, often leasing the land from the government (Rahmato, 2005).

Since the Irist period, a decision on transaction of land was under strict control by the local community. The Irist system is believed to be the first tenure system in the region. The key informants interviewed for the elaboration of this study reported that land was free to everyone before the Irist system. Their argument to support this logic is the literal meaning of the word Irist itself, which means the place to rest. They also argue that Irist holders are the descendants of the first settlers in the area, to whom the right to hold land was given.

3.2.3 Legal Cadastral Domain Model

The relation between land and mankind can be created and developed by formal or informal rules (Williamson, et al., 2010). Informal property rights are those without official recognition of the state. In some cases they can even be in direct violation to the formal rules. In informal property right systems the community defines which land-related activities are permitted and which not. The restrictions in the informal setting are imposed by the local society and enforced by social sanctions (Shibeshi, Fuchs & Mansberger, 2013).

At the FIG Congress in Munich in 2006, it was proposed to develop a (shared)\core cadastral domain model. The recommended model shall serve at least two important goals: (1) avoid reinventing and re-implementing the same functionality over and over again, and (2) enable involved parties, both within one country and between different countries, to communicate based on the shared ontology implied by the model (Lemmen & Van Oosterom, 2006). It is easy to complicate land issues but it is very difficult to describe land issues in a simplified form (Lemmen, 2012).

The fact that land administration systems are dealing with society specific tenure problems makes it difficult to copy a well working system from one country to the other. Therefore, standardization of models is forced to be at conceptual framework level (Ali, Tuladhar & Zevenbergen, 2010; Van der Molen, 2002). Recently, ISO published the international standard on land administration (Land Administration Domain Model / LADM) that can be used for the exchange of knowledge on this topic (ISO, 2012).

After the approval of LADM, studies are going to be used as input for upgrading. An option for expanding was the LADM with legal profiles (Paasch et al., 2013a) and RRR (rights, restrictions and responsibilities) (Paasch et al., 2013b) can be mentioned as examples. The upgrading of LADM requires the description of many land administration systems as input.

The concept of legal cadastral model is based on the relation between humankind to land, which can be classified by beneficial and limiting rights. Two major interests on land are defined in the model: Public interest and private interest on land. The details about the relation and interests on land are managed differently in different jurisdictions.

Combining formal and informal systems will continue to be a challenge (Ubink & Quan, 2008). Contextualization of standard framework models is required to understand properly the two systems (Paasch, 2011; Paasch, 2012). In this thesis, the informal rights were categorized into similar groups. Criteria for the categorization were developed. The core legal cadastral domain model was customized to describe both the formal and the informal holding right system in ANRS.

3.3 Toolbox for Managing Interests on Land

The toolbox to manage interests on land can be used to compare different land administration systems. The toolbox cannot cover all land interests on land. A mix of qualitative and quantitative analysis is required to develop a toolbox that can guide the development of progressive land administration system. The results of case studies are potential sources of information for the development of a toolbox (Bennett et al., 2008).

The RRR toolbox developed by Bennett & Rajabifard (2009) is supposed to manage a myriad of social, economic and environmental reasons that can impact implementation of land administration system. The RRR Toolbox is the result of applied scientific research. It consists of eight principles (Bennett & Rajabifard, 2009). The advantage of toolbox approach is its flexibility to fit in diverse environments. The flexibility and suitability of toolbox approach for diverse environments is also reported for manufacturing industry (Georgoulias et al., 2007).

Tool box approach can be used for different areas. The ITC GEONETCast toolbox is developed to manage spatial data. The approach underlined the importance of changing data streams to information to support informed decision making (Maathuis, Mannaerts & Retsios, 2008). Support for informed decision making and efficient utilization of spatial data are necessary conditions for progressive land administration systems. The toolbox approach can also be used for creative arts (Hanson & Herz, 2011).

The toolbox approach is also recommended for land administration purposes. The land administration toolbox has general tools, professional tools and emerging tools. The general tools include;

- Land policy tools

- Governance and legal framework tools
- Land market tools
- Marine administration tools
- Land-use, land development, and valuation tools
- ICT, SDI, and land information tools
- Capacity and institution-building tools
- Project management monitoring and evaluation tools
- Business models, risk management, and funding tools. The professional tools include;
    - Tenure tools
    - Registration system tools
    - Titling and adjudication tools
    - Land unit tools
    - Boundary tools
    - Cadastral surveying and mapping tools
    - Building title tools. The emerging tools contains the following four sections:
- Pro-poor land management tools
- Non cadastral approaches and tools
- Gender equity tools
- Human-rights tools (Williamson et al., 2010).

3.4 Cadastral and Registration Law

A review of the land registration, cadastral and land information management systems in Africa by Augustinus (2003) indicates that:

- Only a few proportion, less than 1% of Sub-Saharan Africa has developed registration and cadastral system;
- Land information system and GIS is not commonly used by most African countries; and
- Cadastral systems are poorly organized, though are supplying most of the information.

The review showed that the level the status of the cadastral and registration systems in Africa is very law. It is high time to introduce efficient cadastral and registration system in Africa and benefit from globalization. The pre-conditions to consider before installing cadastral and registration system in the developing world are:

- The level of tenure insecurity;
- The status of land related conflicts;
- Early land market development;
- Need for credit base;
- Problems on redistributive land reform (Hanstad, 1997).

Every country requires a proper strategy to manage the relationship between society, people and land. These strategies have to be periodically updated to address the dynamic nature of the relationship and the status of the existing land administration system. An efficient land administration system guarantees sustainable development, whereas the design and the implementation of the land administration infrastructure have to be based on the following principles:

- Land policy principles;
- Land tenure principles;
- Land administration and cadastral principles;
- Institutional principles;
- Spatial data infrastructure principles;
- Technical principles;
- Human resource principles (Bennett & Rajabifard, 2009; Williamson, 2000).

Land administration and cadastral principles have to be defined very detailed and they have to be adjusted to the requirements of the specific country. Based on the above mentioned principles the land administration system has to address private interests, government management interests, and government access interests on land (Bennett et al., 2008). A private interest in Ethiopian condition mainly deals with the RRRs (Rights, Restrictions and Responsibilities) of individuals, state, and communal holdings. The government management interests are related to the state responsibility for maintaining good relationship between society and its land. Government access interests mainly focuses on tools necessary for the management of interests on land. Many of the interests on land are parcel based and therefore can be registered and mapped. Others can be managed by only using legal provisions and policy implementation tools. The plan to upgrade the primary book of holdings into second level book of holdings, including spatial description of holdings in Ethiopia, is part of managing parcel based interests on land.

The implementation of an efficient cadastral and registration proclamation will contribute to guided legal actions, will enable transparent decisions, will guarantee uniformity, will avoid costly surprises, will protect public interests, and will increase the efficient use of land and other natural resources. The guiding principles to be considered during the development of a cadastral and registration proclamation can be defined in a tool.

3.5 Reference Points for Cadastral Surveying

A geodetic reference network is a set of geodetic control points: physical monuments and point descriptions that contain known geodetic coordinates of latitude, longitude, and elevation (IAAO, 2004). According to the Ethiopian land law all cadastral maps have to be connected to the national grid (FDRE, 2005). The importance of connecting cadastral maps as well as the importance of network accuracy is also documented in literature (e.g. Larsson, 1991; Craig & Wahl, 2003).

Reference framework has to be established before the commencement of survey projects. In other words, the first step in survey projects is to create benchmarks that are connected to the national grid (Fradkin & Doytsher, 2002). The level and accuracy of geodetic control points can be different depending on the intended use and accuracy of the survey project. But the general rule is the quality and accuracy of the control points must be better than the most accurate need of the intended surveys (Fradkin & Doytsher, 2002, Popovas, D. 2001).

The two basic types of geodetic controls, horizontal and vertical, are commonly used in cadastral surveys. The horizontal geodetic control data consist of distances, directions, and angles between control stations. This data is used to determine geodetic coordinates and azimuths. The geodetic coordinates (latitude and longitude) can be converted to other coordinate systems.

The vertical control networks have been established to provide a means of referencing heights of stations above a specified surface. The height is measured along the direction of the plumb line between the point and the reference surface. The reference surface is the geoid, which closely approximates mean sea level (Bedada, 2010).

The level of the accuracy of geodetic control points is usually expressed in terms of ratios. The commonly used standard is shown in *Table 1* below FGCC (1984).

Table 1: Standard of geodetic control points (source FGCC, 1984)

no	Classification	Minimum distance accuracy
1	A-Order	1:10,000,000
2	B-order	1:1,000,000
3	First-order	1:100,000
4	Second-order, class I	1:50,000
5	Second-order, class II	1:20,000
6	Third-order, class I	1:10,000
7	Third-order, class II	1:5,000

In order to conduct different kinds of surveying and mapping a common reference framework of control points are required. The cadastral surveys that are supposed to answer the question where and how much needs to be connected to the national grid. Plans that are not connected to the national grid can hardly be used to locate properties using coordinates (Larsson, 1991, Williamson, 1983).

Ethiopian mapping Agency (EMA) is the responsible authority for mapping and establishing geodetic points in Ethiopia. The Ethiopian Mapping Agency is using Clarke 1880 spheroid and UTM projection in metric unit based on Adindan Datum, also sometimes called the Blue Nile datum. Until now, about 80% of the country is covered with primary and secondary geodetic control points with an approximate interval of 50 km.

The Blue Nile datum of 1958 created with the support of U.S. Department of commerce coast and Geodetic survey for Ethiopia was the first datum of Ethiopia (BNBSP, 1961). The origin of the geodetic work was in southern Egypt, south of Lake Nasser, at station Adindan where $\Phi_0 = 22°10'07.1098''N$, $\Lambda_0 = 31°29'21.6079''$ East of Greenwich, the deflection of the vertical $\zeta = +2.38''$ and $\eta = -2.51''$, and the ellipsoid of reference was the Clarke 1880 (modified) where a = 6,378,249.145 m and $1/f = 293.465$. The Ethiopian Transverse Mercator grid is based on a central meridian where $\lambda_0 = 37°30'E$, scale factor at origin where $m_0 = 0.9995$ (Blackwell, 1962; Mugnier, 2003).

Ethiopian mapping agency (EMA) has a plan to density and upgrade the geodetic points all over the country (Mugnier, 2003). Compared to the demand, the densification process of EMA is very much lagging behind.

The plan of EMA shows that the future focus will be on the densification CORS (Continuously Operating Reference Station) and first order points all over the country. But even after the construction of all the planned points, the density of EMA control points will be very far from enough for undertaking cadastral surveying that requires on average one point in every kilometer square (Sultan, 2011). *Figure 1* shows the distribution of both planned and existing EMA points in the study area.

Figure 1: Planned and existing ground control point distribution in the study area (source; based on Sultan, 2012)

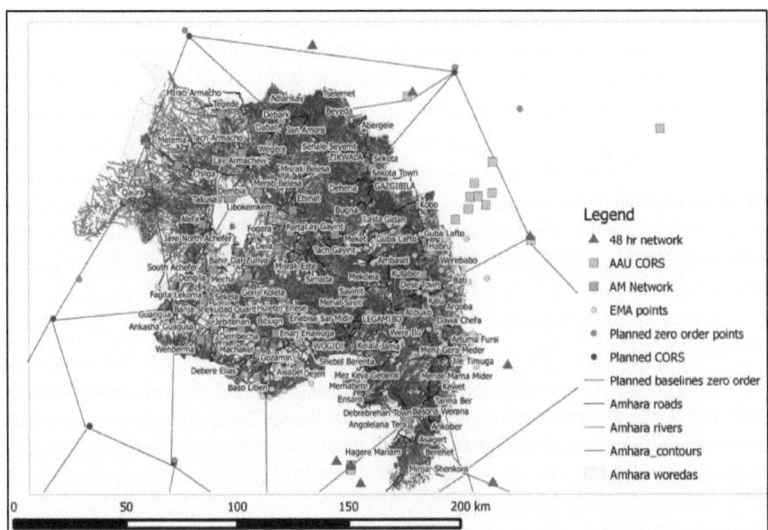

BoEPLAU took the initiative to create 42 geodetic points distributed over the entire ANRS and to connect them to the national grid. EMA was asked to evaluate the quality of the control points and to consider them as EMA points after approval. Unfortunately, the approval process took very long and became subject for controversy on mandate. As a result, the new established control points are not recognized officially by EMA until yet.

AM network is established by BoEPLAU and used as a first test for the planned second level certification program. The network includes 42 points. Two of the points of AM network (Gondar and Debanka) are on the same sites as EMA airport points. All control points of AM network were adjusted and connected to the national grid (Miskas & Molnar, 2010). The planned second level certification program and cadastral mapping in ANRS will be connected to geodetic control points.

3.6 Selection of Cadastral Surveying Methods in Progressive Land Administration Systems

3.6.1 Review of experiences and practices

In many countries the legal rights might be clear but the spatial extent and location of the 'parcels' of property over which these rights apply may be unclear (Barnes, 1990). This is the case in many deeds registration systems that index information by 'owner' or 'document' rather than by 'parcel' such as the traditional public registries in much of Latin America.

One of the dangers of increasing accuracy and decreasing costs is the specification of a standard just because it is technically possible rather than because it is needed.

A lot of the survey inaccuracy in many countries is due to a lack of understanding of survey and measurement methodology, accuracy, precision and error theory. In many countries there is also confusion regarding coordinate datum, geodetic control and issues related to reference frames Ethiopia included.

No project in the developing world has been able to implement and sustain high-accuracy surveys over extensive areas of their jurisdiction. Those countries that have been successful in registering significant numbers of titles have tended to concentrate on relatively simple, low cost survey methods and produced graphical standard cadastral index maps.

3.6.2 Selection of Cadastral Methods

The proportion of the Earth mapped at a scale of at least 1:50, 000 is 60 percent. The proportion for a scale that is suitable for land ownership (1:2,500 – 1:10,000) is much smaller than that (Corlazzoli & Fernandez, 2004). Large scale mapping and well-functioning cadastral systems are confined to the developed world.

Survey and mapping technology has improved greatly in recent decades. The use of digital orthophotos produced from aerial images and from high-resolution satellite imagery is very cost-effective and has been applied for mapping in many cadastral projects throughout the world. Automated parcel map making from HRSI by using segmentation programs found to be of little help because of the nature of boundary lines (Haile, 2005). Boundary lines are not always visible and therefore they cannot be easily identified automatically by segmentation software. However, the method can still be useful, if boundary identification is supported by fieldwork.

GPS equipment and electronic total stations have increased in accuracy and flexibility and decreased in cost and have been used on many projects throughout the world. However, even with reduced costs base mapping and ground survey can be as much as 50% of the cost of a major project systematic registration project (Burns, 2007). But more importantly the higher the accuracy specified for boundary surveys, the higher the cost in terms of equipment and skilled operators necessary for the on-going maintenance of the spatial framework that supports the system to record property rights.

Many countries have specified too high accuracies for surveys and often these high standards are poorly enforced. Fit to the purpose land administration approach is recently developed by joint effort of FIG and the World Bank. The major elements included in the approach are: flexible, inclusive, participatory, affordable, reliable, attainable and upgradable. The basic components of the fit to the purpose concept are using affordable modern technologies for building spatial infrastructure, participatory nature and adopting the legal framework that can foster flexibility (Enemark et al., 2014).

The over ambitious plans by some land administration projects regarding early establishment of spatial infrastructure with in short period of time all over the project area is more a problem than a solution for the establishment of progressive land administration systems. The inclusion of surveying and mapping component in almost every new land administration project is the indication of the strong need to have a well-functioning cadastral system. On the other hand, the spatial infrastructure is the most demanding and complex part of a land administration system (Bromley, 2008).

Land rights in Ethiopia were poorly defined and used to be the major cause of tenure insecurity until the establishment of the new land administration system (Rahmato, 2005; EEA, 2002; Mesfin, 1991; Deininger, et al., 2003; Nega, Adenew & Gebre Sellasie, 2003; Haile 2005). After the establishment of progressive land administration system in ANRS, the tenure security feeling is improved and land rights are better defined (Alemu, 2012; Ambaye, 2013; Deininger, Daneal & Tilahun, 2011; Deininger et al., 2008).

Pilot projects and trials were conducted in ANRS to choose the suitable surveying and mapping technique for cadastral projects in Ethiopia. Pilot projects and trials were limited to searching specific solutions for defined problems in a specific project context (Shibeshi, Fuchs & Mansberger, 2013).

Some of the examples of donor supported pilot projects are Adisen Gulit (SIDA), ELAP, ELTAP (USAID) and Angot yedegera (FINIDA and WB). The World Bank and FINIDA supported pilot projects are focusing on the use of remote sensing tools for cadastral purposes. The support from SIDA and USAID is focused on ground survey methods (Shibeshi, Fuchs & Mansberger, 2013).

4 Study Area, and Methods

4.1 Study Area

Ethiopia – officially called The Federal Democratic Republic of Ethiopia, in short FDRE – is an country in East Africa. The country covers a total area of 1.1 million km² and has a total population of more than 85 million. Ethiopia is divided in to nine regional states. The current study was outlined in one of the nine regional states, namely in the Amhara National Regional State (ANRS). ANRS is located in the north-western part of Ethiopia between 9°45'N and 13°45'N, and 35°15'E and 40°15'E. The region has an area of 154,708 km² and 18 million inhabitants. A total of more than 3.6 million land holders are registered in ANRS (CSA, 2007). (See *Figure 2*, Location of ANRS).

Figure 2: Location of ANRS

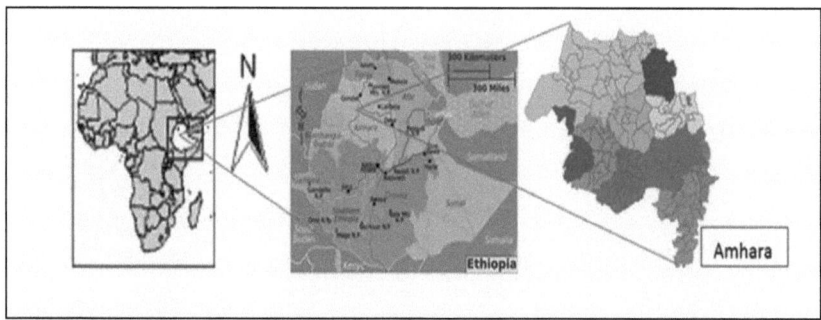

ANRS is the pioneer to start the implementation of a land administration system. The study area was chosen, because the cadastral mapping and the issuance of second level book of holding will be launched as a pilot in ANRS – according to the country's five years growth and transformation plan.

4.2 Methods

The study deploys both quantitative and qualitative data collection methods as the major data source for developing an evaluation framework that can help to draw lessons from progressive land administration systems.

The methods used were literature review, individual interviews with individual farmers and with land administration committee members, professionals' expert panels with Woreda (district) and zonal expert. Group discussions with major stakeholders and discussions with key informants were carried out. The discussions were based on the results of a questionnaire sent to all land administration offices in the region. In addition, long-term static measurements on sample geodetic control points and comparison measurements using different survey methods were outlined.

The desk work includes the evaluation of legal and policy documents, the evaluation of different official reports, law and policy documents, publications and relevant scientific theses. The results of the literature review were used to guide the need assessment part of the study. Evaluation and description of ANRS rural land administration system was conducted by using data from need assessment stage. Field data collection for the comparison of suitable methods was based on the results of the need assessment stage. The design in *Figure 3* shows the flow of data collection and the link between different survey tools.

The data collection was conducted between June 2011 and July 2013. The field data were analyzed using descriptive statistics and content analysis methods. The author has collected data from both, primary and secondary sources.

Figure 3: Research design

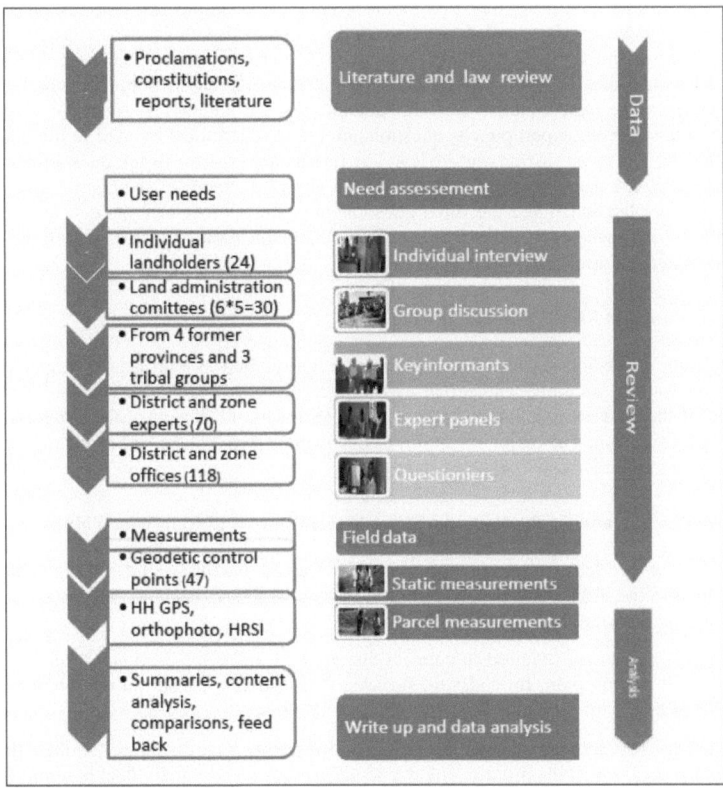

4.2.1 Individual Interviews

Individual interviews were used to comprehend the level of the satisfaction of system users. The individual interviews were conducted with small scale land holders and with land administration and land use committee members.

The aim of individual interviews was to fathom the effectiveness of ANRS rural land administration at policy, management, and operational level indirectly by the level of customer satisfaction. Semi structured interviews (Simon 2006) focusing on the major interests small scale (household) farmers were used to evaluate their level of satisfaction on land administration activities. 24 farmers from six different Woredas (administrative structure with the level of district) were interviewed. Additionally, five members of land administration committees from each selected Woreda were interviewed (in sum 30). The sample of individual interviews had determined randomly selected taking into account to get a proportional number of female respondents.

4.2.2 Questionnaire Survey and Expert Panels

Expert panels are discussion forums with district and zonal experts as groups. The experts are working on rural land administration; this includes land use planning and land valuation. Experts' panels guided by open ended questions were conducted with 15 expert groups in seven zonal offices and in eight districts. A total of 70 experts attended the panels. The discussion with the professionals focused on the accomplishments, on the bottlenecks and on recommendations. The expert panels were all attended by the author. To supplement the information obtained from expert panels, questionnaires were distributed by mail to the land administration offices of all districts in ANRS. The instruction how to fill the questionnaire was also attached. The questionnaires were filled by a group of experts in the district office. The meetings were facilitated and the filled questionnaire was approved by the office head. 118 institutions responded to questions focused on four main topics (general issues, land tenure, land value, land use).

4.2.3 Discussion with Major Stakeholders

The expectations of major stakeholders (revenue authority, bureau of justice, bureau of agriculture, investment authority and urban and industry development) were assessed. The contribution of the land administration system in ANRS for the fulfilment of their missions was a point of discussion. The major constraint of this method was that the stakeholders were evaluating the system only based on their perspective.

4.2.4 Identification and Evaluation and Static Measurements of Reference Points

The suitability of geodetic reference points was evaluated based on the requirements. Requirements for the Ethiopian Ground Control Network concerning the Second Level Certification Program (defined by BoEPLAUA) are:

- Reference points are required to connect the cadastral maps to a national grid. They serve as a starting point for cadastral surveys. A distribution density of one point per km^2 is desirable.

- Accuracy: The accuracy of the geodetic control points have to be better than the intended accuracy of the project. The intended accuracy level of individual holdings in the forthcoming second level certification program is ±0.3 m in plane. As boundary points only will be assessed in plane, the accuracy for heights is not defined in the cadaster project. But the reference points also should be used for other surveying projects. So the accuracy needs are more rigorous for these projects. Therefore BoEPLAUA demands an accuracy of ±0.08 m in plane and ± 0.10 m in height.

- Stabilization and Signalization: The need for inter-visibility between geodetic control points is very much reduced after the introduction of GNSS equipment. Open sky or sky visibility and reduced multipath are a critical requirement during point establishment. However visibility is still important for total station surveys. The requirement during point establishment is therefore sky visibility up to 10° cut off angle and horizontal visibility of up to 300 meters at least in three directions.

- Documentation: Documentation shall start with an unambiguous name and identifier for the point. The point location descriptions the coordinates of the point are the most important information to be documented properly.

In the study, the usability of the existing Ethiopian control point network is investigated regarding the following criteria:

- Discoverability: It is proved, if the site description is suitable for locating easily the control points.
- Availability and status of control points: It is checked, if the monument is still available and accessible.
- Accuracy of control points: It is checked, whether the accuracy requirements defined by the project manager of the second level certification program can be met.

All 42 control points of the AM network were investigated. Due to the high number of control points, the observation of all EMA points was not possible. In sum 120 control points were purchased by EPLAUA from EMA for cadastral purpose in ANRS. 30 points were selected randomly by taking every fourth out of the whole list of points with names ordered alphabetically.

Availability is described by the accessibility and the condition of the point. The possibility of mounting an instrument on the point has to be easily possible.

The site description of both, AM and EMA control points, was investigated. The quality of the template, the practice and the usefulness of the description of all AM points and the selected EMA points were considered. The identified needs are to be unique, systematic, easy to remember and explanatory. The point description to be included in the documentation should contain textual and graphic descriptions and sketches that can easily help to identify the point. The documentation needs to be easily available to all users.

The AM network is connected to Addis and Jima core stations (www.ipg.tu-darmstadt.de/en/adis_igs.html). The data from the two stations was converted to a common epoch using a velocity factor. It was also changed to common reference frame using seven predefined transformation parameters. The velocity factor was generated using NUVEL 1A model for plate motion calculator developed by UNAVCO was used to adjust the coordinates of control stations in Addis and Jimma.

During the field work, static measurements were conducted on 96 AM points and on 7 EMA ground control points using dual frequency Leica 900 GPS. The precise orbits and clocks used by Canadian Spatial Reference System - Precise Point Positioning (CSCR-PPP) remove a large part of the Global Navigation Satellite System (GNSS) errors. In addition, CSCR-PPP processing must also properly account for several other effects on the position of the GNSS receiver (Moreno et al., 2011; Seredovich, Irughe & Ehigiator, 2012). The known coordinates of control points were compared with the result of CSRS-PPP and cm level standard errors both in North and East were recorded. Other similar studies also confirmed mm to cm level results at 99% confidence level in different places (Chris et al., 2012; Ebner & Featherstone, 2008; El-Mowafy, 2011; Seredovich, Irughe & Ehigiator, 2012).

For the accuracy assessment of EMA and AM control points the existing coordinates were compared with the static measurement results of GNSS methods. The static measurements were conducted for long-term, 4 to 8 hours, depending on the distance from the base line, on the number of visible satellites and on the DOP (Dilution of Precision) values. Precise ephemeris was calculated using data provided by the International Ground Station (IGS). The base lines were created in a rectangular pattern combining two known and to unknown points at the same time. The equipment used for the static measurement was Leica 900 and the antenna type used was ATX 900. The result was post processed using Leica geo office software and all data was converted to RINEX format (Receiver Independent Exchange format) that can be submitted to CSR-PPP to get calculated results.

The static measurement data of both EMA and AM sample points is converted to RINEX format and uploaded on the web pages of CSCR-PPP, NASA-APPS and AUPOS – PPP (AUSPOS, 2012; NASA, 2012; NRCANGSD, 2012). The East and North coordinates of the sample points were converted to common local datum, Adindan Datum, and then the results were compared with each other.

The regression test is used to understand the relationship between the East and North coordinates of AM and EMA networks. The relationship is represented by correlation coefficient (r). The value of the correlation coefficient is always between -1 and +1. One means perfect linear relationship between variables. The two tailed Pearson correlation method at 0.01 levels was used to test the relationship between measurements. The assumptions for Pearson's coefficient are data in an interval scale and normally distributed scores.

4.2.5 Comparisons of Survey Methods Using Sample Measurements

The case study is a recommended method for most cadastral studies. A case study is suitable because it facilitates in-depth and holistic investigation of cadastral systems. The research using a case study method can be a positivist, interpretive or critical (Zevenbergen, 2002; Williamson & Ting, 2001; Ali, Tuladhar & Zevenbergen, 2010).

For the current thesis, exploratory case study is used to compare the needs and the capabilities of the different mapping methods. Remote sensing (Orthophoto and HRSI) and ground survey (GNSS and total stations) tools were compared. The sample tools were selected for the reason that they are practiced in ANRS. Before technical comparison for the cadastral survey tools is made, the systems capabilities to define proper objectives, the suitability of existing way of working and timing of the intervention were evaluated.

The methods used to define the objectives and aims were policy and law review, observations of the current practice, literature reviews and experiential learning from the case study of ANRS. Similar methods were used to investigate timing and need assessment criteria. The focus of correct timing and need assessment was on understanding the readiness of the users, stakeholders, the institution policy and law and so on. During the field work the readiness was assessed by group discussion, individual interview, expert panels and questioner survey.

The correct way of working was identified mainly by reviewing the field guides and procedures used in different pilot programs. The major sources of information were the way of working developed by BoEPLAU, the procedure for the implementation of ELTAP and ELAP, and the guideline developed by RELA for the implementation of pilot projects on surveying and mapping. The decision criteria deals with the necessary considerations to develop a working procedure suitable for the implementation of cadastral projects in progressive land administration systems.

The development of a toolbox to guide the selection of proper techniques and technologies suitable for cadastral systems is a dynamic and ever changing process caused by changing cadastral systems and rapid development of measuring technology. The current study categorized the need for cadastral systems into two major groups: The need during the establishment of the cadastral systems and the need for maintenance and updating. Demands on accuracy are also dependent on the holding type, the value of the property, etc. Therefore the criteria cannot be the same for different scenarios. The methodology used in this study is first to create major and sub categories that can address the major scenarios. Techniques and technologies are also grouped in to two major groups; namely ground survey tools and remote sensing tools.

The commonly used techniques from remote sensing and ground survey tools (orthophotos and high resolution satellite imagery from remote sensing tools and hand held GPS from the ground surveys) were tested for accuracy, cost and speed. RTK GPS surveys were carried out to control point accuracy of tools mentioned above. The accuracy test is conducted following the measurement practice during piloting. Cost and speed estimates are based on data analysis during sample pilot projects.

Remote sensing tools (orthophoto produced from aerial photos with 0.5 cm ground resolution and HRSI from Quickbird and WorldView 2) and hand held GNSS surveys from ground survey tools were compared with survey data from RTK GNSS. The two sites for HRSI comparisons were Angot Yedegera (629 point measurements) and Zenbela (213 point measurements). The data for Anagot Yedegra was a primary data while the data for Zenbela was a secondary data collected by Andent. The total of 842 point measurements were used to compare RTK GNSS and HRSI measurements.

The orthophoto field test was conducted in Seraba Parish. Randomly distributed points (1098 in number) were measured for comparison with RTK GPS measured equivalents. The field test for the comparison of hand held GPS and RTK GPS measurements was conducted in Angot Yedegera (287 points) and Illu parish in Oromea region (56 points). The secondary data of Andent in Zenbela parish (93 points) was also considered. The total of 436 point measurements in three sites were used for HH GPS and RTK GNSS comparisons.

The data from the pilot project reports of ELAP (USAID supported project) and RIELA (FINIDA supported project) are used for speed and cost estimate comparison. The human power requirement is to estimate the total number of personnel required to establish spatial data infrastructure. The human power for establishment is dependent on the type of methods to be deployed. To compare and to identify the most proper methods, the number of parcels being surveyed per day per surveyor was used as an indicator.

4.2.6 Compiling and Processing of Findings

Findings from the review of official documents were linked with the results of individual interviews. The results obtained through questionnaire surveys and interviews were commented during expert panels. The findings were presented during wrap up meeting with regional experts and representatives of major stakeholders and checked for validity.

The results of all methods were brought into a proper digital format. For the survey qualitative as well as quantitative data were collected. Methods of descriptive statistic were used for the quantitative data analysis. Content analysis according to Mayoux (2006) was the method outlined to analyze data from qualitative questionnaires and discussions.

Triangulation methods were applied to compare and validate data from surveys of qualitative and quantitative methods (Mathison, 1988).

5 Results and Discussion

5.1 Evaluation Result of Land Administration System in ANRS

The ANRS is implementing progressive land administration system since 2003. But no systematic evaluation was conducted since then. An evaluation allows the comparisons of what is planned and what is performed at the system level.

The evaluation outlined in the current thesis identified the most important tasks that have to be addressed before implementing the planned second level certification program: Development of methods for the institutional set-up, for the legal system, for the densification of ground control points, and for the cadastral survey.

The evaluation covered the status, the pros and the cons of ANRS rural land administration system at policy level, management level, operational level, external factors and review processes following the recommendation of (Steudler et al., 2004). The customized evaluation framework was applied using policy review and discussions with major users and implementers of the system.

Figure 4: Elements of Evaluation framework

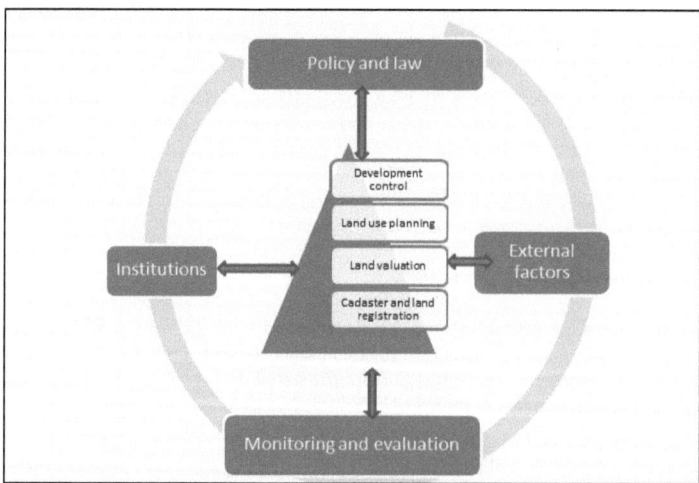

The land administration professionals in ANRS were asked to discuss and rate the system. The overall rating of the system during experts' panel was very good (4.2).The summery result are documented in *Table 2*. Similar results were confirmed by individual interview with randomly selected land holders. The evaluation framework includes the evaluation of the status of policy and law aspects, the effectiveness of involved institutions, the implementation status of core land administration functions, the influence of external factors and the status of inbuilt monitoring and evaluation mechanisms (see *Figure 4*).

Table 2: Summery questions rated from 5 (excellent) to 1 (poor) based on respondent's experience. (Number of samples: 15 groups with total attendants of 70)

#	Question																Average	± σ
1	Do you evaluate the land administration system in your area as a successful system?	3	4	3	5	5	5	4	5	3	4	4	5	4	5	4	4.2	0.77
2	How important is land administration system for reduction of land related conflicts?	4	3	5	5	5	5	5	5	5	5	5	5	5	5	5	4.8	0.56
3	Are landholders in your area willing to recover all costs of efficient land administration services?	5	5	5	4	5	4	4	4	5	4	5	4	5	5	4	4.5	0.51
4	How impermanent is the spatial data set for land administration in your area?	4	5	4	5	5	5	5	5	5	5	5	5	5	5	4	4.8	0.41

5.1.1 Policy and Law

The land administration system of ANRS is guided by constitutional provisions, land administration and use policy of ANRS, pertinent proclamations, regulations and directives. In Ethiopia land policy formulation is a responsibility of the federal state. The most important land administration issues are incorporated in the constitution of the country. Federal land administration framework law was first enacted in 1997 FDRE (1997) and later amended in 2005 (FDRE, 2005).

The Constitution of the Federal Democratic Republic of Ethiopia has enshrined the basic principles about the property right of citizens under (FDRE, 1995) (Article 40. Sub-article 1). This article generally provides that *'Every Ethiopian citizen has the right to the ownership of private property. Unless prescribed otherwise by law on account of public interest, this right shall include the right to acquire, to use and, in a manner compatible with the rights of other citizens, to dispose of such property by sell or bequest or to transfer it otherwise.'* From the reading of this article one can assume individualized property right system in Ethiopia. But property in the context of the constitution is not including land. Therefore transfer rights given by this article are excluding land though in practice the transfer of fixed assets is including the parcel they are built on it.

Land without any fixed property on it, is not subject to sell and the issue is proclaimed as: *'The right to ownership of rural and urban land, as well as of all natural resources, is exclusively vested in the state and in the peoples of Ethiopia. Land is the common property of the Nations, Nationalities and Peoples of Ethiopia and shall not be subject to sell or other means of exchange'* (FDRE, 1995). The objective of this article is said to be the protection of peasant farmers and pastoralists from eviction. But when the land is needed for public purposes the state has the power to expropriate.

The provisions in the constitution protect the landholders from dislocations. The landholders shall never lose their occupation without proper compensation. The constitution states the subject in point as: *'Ethiopian peasants have the right to obtain land without payment and the protection against eviction from their possession...'* (FDRE, 1995). However the payable amount is not considered fair and many application related problems that can affect the security feelings of land holders are reported (Ambaye, 2013).

Holding right includes one or more parcels and a parcel is the smallest spatial unit. The right to hold property is also stated in the constitution Article 40/7. This right is given for every Ethiopian and the protection includes all immovable improvements made by the citizen. The improvements can be caused by peoples' labor, creativity or capital inputs on land. Their rights include the right to alienate, to bequeath, to transfer and to remove their property when the right to use the land expires.

The land administration system of ANRS benefits from the significant status given to land issues in the constitution. The objective of giving land issues a relative permanence is to protect core land related functions from frequent changes, caused by political turmoil and/or other external factors, before the land policy bear fruits – normally a process that takes a longer period of time. The argument was supported by respondents (land administration professionals).

The constitution is the highest law in the country addressing the most important issues and principles. Therefore according to respondent professionals, land has to be a central part in the constitution. The other objective of the constitution is to create and maintain a nation. Land is identified by respondents as a main factor to create a unified economic, social, and political entity. The constitution can be said incomplete if it lacks land issues in it.

On the contrary some people argue that placing dynamic land issues in the 'static' constitution is the wrong move. According to them, flexibility is required to manage effectively land issues, which is hardly possible for constitutional matters (Adal, 2002; Rahmato, 2005). The points raised in the argument are valid but we have to choose either the need for permanence and consistency or the ability to entertain dynamism. The respondents reported in favor of permanence than dynamism for framework land policy issues and flexibility and dynamism while practicing detailed activities based on constitutional provisions.

The federal framework land law, which is based on the constitution, defines the holding right as the right of any peasant farmer or semi pastoralist and pastoralist:

- to use the rural land for the purpose of agriculture and natural resource development;
- to lease and to bequeath land to members of their family or other lawful heirs;
- to acquire property produced on their land thereon by their labour or capital; and
- to sell, exchange and bequeath same (FDRE, 2005).

The framework federal law gave to national regional states the right to enact land administration proclamation. Based on the federal proclamation and their constitutional right the four main national regional states enacted their regional land laws.

The ANRS tested different land administration activities on two pilot projects before starting large scale implementation. The way of working was one of the most important outputs of the two pilot projects in ANRS. The mission, the vision, the strategy and the measures for implementing a land administration system in ANRS was developed based on the experiences and lessons from the two pilot projects.

The scale, pace and cost-effectiveness of ANRS rural land administration system was unprecedented in Africa and in a participatory and public process (Deininger et al., 2008). In more recent literature positive welfare impacts (Holden & Ghebru, 2013) and increased investment in terms of land productivity and land rental market activity was reported (Bezabih, Holden & Mannberg, 2012; Deininger, Daniel & Tilahun, 2011; Holden, Deininger & Ghebru, 2009, 2011). The participatory way of working developed for guiding the implementation of the system enables the response to the needs of society, especially to protect the weaker parties in the society.

The results of the panel discussion with Woreda and Zonal land administration professionals confirmed that the knowledge about the mission and vision of land administration system of ANRS is good enough. But no respondent group had correct understanding and background knowledge on the strategy document. The objectives and tasks of ANRS land administration system are well defined by legal statues at different level, and they were correctly articulated in all fifteen panel discussions.

The status of the implementation of the general issues in ANRS rural land administration system was rated high. The policy formulation and monitoring, legal system implementation and gender issue were the activities performed to the satisfaction of the professionals. Research and development, computerization and IT, public information and awareness, capacity building, self-financing and cost recovery activities were rated low.

Past experience of landholders about government intervention on land issues has not been positive. Frequently, government funded land redistribution decisions were taken without the consent of landholders, and land used by landholders was dedicated to public purposes without any court decision or any compensation payment. These negative experiences were reported to pose a big challenge on the current land administration project. Many information campaigns and numerous awareness creation programs about the new land policy and the adapted legal system were conducted to minimize this problem. The increasing value of the book of holding to win court cases has significant contribution to gain trust.

Nowadays tenure security feeling – one prime objective of the implementation of land administration system in ANRS – is increasing. Long term investment and natural resource conservation measures are reported to be improving. Land fragmentation and shortage of farm land continued to be critical challenges to improve the income level of the small scale farmers. According to individual farmers respondents the average holding size in the region is 1.1 ha per household and the average number of parcels per landholding was reported to be five. The average size of one parcel is reported to be less than 0.25 ha. Some even use the term starvation plots to indicate the small size of parcels in the area (Rahmato, 2005). The fragmentation is mainly the result of frequent land redistribution and an attempt to get equitable share from different land types.

According to the land law of ANRS land consolidation is voluntary. The land holders are encouraged to amalgamate the parcels in their holdings. The objective of land consolidation in ANRS is to increase the efficiency of small scale farmers by reducing the travel time between parcels and cost of management. On the contrary, having widely dispersed parcels is reported as advantage to prevent total loss during natural calamities, such as flood and drought.

The study results show that land administration system in ANRS is participatory and operating properly. Landholders participate in the land administration processes both directly and through their representatives. The representatives of land holders in a parish, called land administration and land use committees, are selected by people and serve for a term of two years. The committee members can be re-elected, if they get majority support for the next term. Among the committee members at least two members have to be females. The direct participation is enabled through public hearings.

Land holders living in a parish have participate in the public hearings. Public hearings are used as final collective approval meetings by landholders. The data and information collected by land administration committee members will be presented to all land holders in the parish and will get approval. The collected data will be posted in public for at least two weeks before the public hearing date.

Approved data and information will be given as a temporary certificate for each landholder and wait for at least one year before final approval and registration. During this time individuals have the right to appeal even on issues approved by the public hearing, as long as they can present evidences. They can also request the district authorities repeat the public hearing based on newly identified evidence and claims. At this stage the role of the land administration and land use experts is only facilitation.

Shemaglewoch shengo (customary land disputes arbitration committee) is another participatory mechanism to manage conflicts at the local level and to reduce conflicts to be presented in courts. The Shemagelewoch shengo is a bridge between formal and customary system. The decisions made under customary rules are acknowledged by the formal system as long as the case is not of criminal nature.

Respondents from major users of the system rated the status of major policy and cross cutting issues. Summarized results of the rating are presented in *Table 3* below. Very little satisfaction is reported on research support. The rate for attention given to computerization and IT is also low. Users are better satisfied with legal system review and monitoring of the planned activities.

Table 3: Summery rating of policy and cross cutting issues from questioner survey (number of samples 118 offices)

No	Issue category and max rating	mean rate
		5
1	Policy formulation and monotoring	3.93
2	Legal system implimentation	3.41
3	Computerization and IT Systems	1.95
4	Capacity building	2.02
5	Public information and awareness	2.38
6	self financing and cost recovery	2.53
7	Research and development	1.88
8	Gender issues	3.70
	Ave	2.725
	standard deviation	0.832

5.1.2 Institution and Management

Institutions are the major tools to transform the legal and policy framework into action. Institutional setting therefore is one of the most important factors contributing to the success of land administration systems. In the division of power between the federal government and regional states, the power to legislate enabling laws concerning the use and conservation of land is vested in the federal government pursuant to Article 51(5) of the federal Constitution (FDRE, 2005). Regional States are vested with the power to administrate land and other natural resources in accordance with federal laws as provided under article 52(2)(d) of the Federal Constitution. The boundary of the responsibilities between federal and regional institutions needs to be clarified by law. The institutional mandate shall emanate from legal provisions.

The Ethiopian mapping agency is responsible for country wide topographic mapping and for the implementation and maintenance of geodetic control points. The responsibility of rural land administration in the ministry of agriculture lies in the responsibility of two directorates. The first is responsible for land use and tenure while the other is responsible for federal level management and lease of state holdings. The two directorates are dedicated to two different state ministers – the reason for this could not be clarified in the study.

In Ethiopia the responsibility to administer and manage land and natural resources was given to the regional states. The regional implementing agencies for rural land administration are varying between regional states – in naming and organizational setting. The Oromia region took a more advanced stage and organized a single bureau responsible for both rural and urban land administration. But after two years the rural and urban land administration responsibilities were separated due to lack of capacity. The Tigray region created a single authority under the bureau of agriculture. The southern region created a core business process under the bureau of agriculture. The other regions followed the example of one of the mentioned regions. In all cases the smallest functional unit is situated at Kebele (equivalent to parish) level.

ANRS enacted the first land administration and land use proclamation in 2000 (ANRS, 2000) and the amendment law in 2006 (ANRS, 2006). The establishment of an authority (EPLAUA) and upgrading it to a bureau, the Bureau of Environmental Protection, Land Administration and Use (BoEPLAU), was based on the regional land law. Another regional decision at management level concerns to give the responsibility to design and to implement the four core land administration functions (land tenure, land use planning, land development control, and land valuation) to BoEPLAU.

The Bureau of Environmental Protection, Land Administration and Use (which is the focus of this evaluation) has three core and nine supportive core processes. The three core processes are environmental protection and sustainability, land administration and land use, and public relations.

The institutional setting of ANRS is a unitary structure addressing all land issues in one institution. The activities of BoEPLAU are guided by a well-articulated mission, vision and strategy. The strategy is designed on the principle of a step by step approach. Complex and resource demanding activities, such as cadastral surveying, are left for later stages. Broadly the strategy can be categorized into four major classes:

- Adjudication, primary book of holding and associated activities;
- Computerization and related activities;
- Cadastral surveys and second level certification;
- Land use planning and development control.

The current organizational structure can be optimized and improved by considering the business processes for each major land administration function. The responsibilities given to some of the existing positions are not clear. For example, the duties of land administration and land registration experts are defined very vague. The assignment of special experts is also difficult to manage. In the process of land administration and land use currently 37 professionals are involved, being responsible for land registration, cadaster, land valuation, state land administration, and land use planning. The number of professionals to be managed under one core process is to high.

Many of the stated weak points of the existing institutional set up were discussed and agreed during the expert's panels. The committee established for business process re-engineering identified similar shortcomings of the current institutional set up. The proposed institutional setup assumes to solve the problems encountered and facilitate efficient management flow. The proposal was presented to BoEPLAU staff and officials. The proposal was also published in one of the main local newspaper to initiate discussion and get feedback (The Ethiopian reporter, http://www.ethiopianreporter.com/component/content/article/309-my-say/72622012-07-28-12-04-24.html). After getting feedback, the agreement on the modified institutional setup was achieved with BoEPLAU officials. *Figure 5* outlines this proposed institutional setup.

The strategic approach of ANRS rural land administration system can be characterized by targeting and achieving small and consecutive wins. This study identified that mission, vision, strategy and the way of working (measures) are appropriately crafted. Only the communication and with it the knowledge of the professionals on the strategy document was a weak-point. This fact is caused by a fast staff turnover as reported by the land administration managers. To avoid the gap in knowledge frequent training and information workshops are required so that the staff can work with ends in mind.

In the whole of ANRS 10 Zonal and 128 Woreda (district) offices are established. At the Kebele level one expert is responsible for all issues related to land. The Kebele expert has to work with the land administration committees at Kebele and at sub-Kebele level. Sub-Kebele committees are selected by the community and amongst the selected members two of them are appointed as representative members of the Kebele level land administration committee.

Figure 5: Proposed organogram

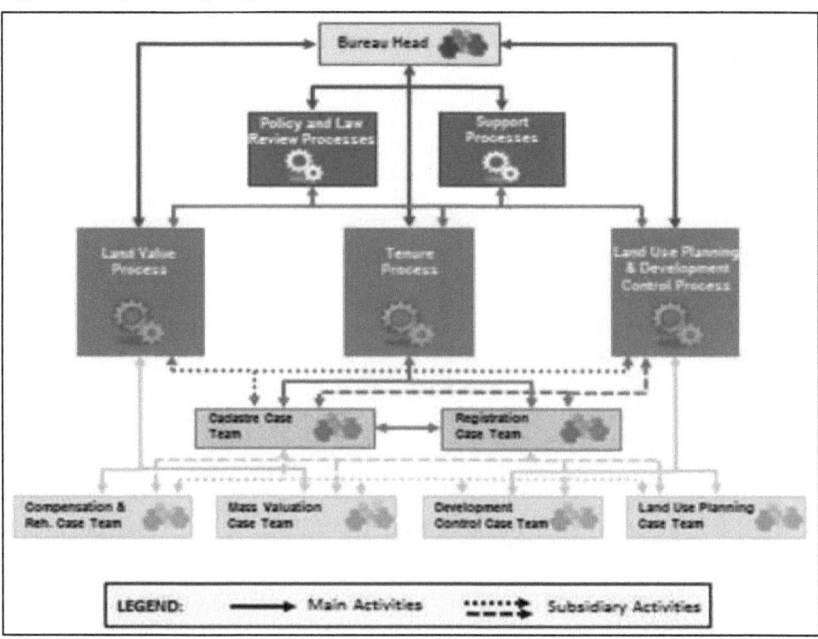

A shemaglewoch shengo (traditional arbitration committee) is established to handle land related conflicts using traditional rules. The participation of the land holders in conflict resolution is enable through the involvement of the elected shemagelewoch shengo. The committee plays a key role in integrating customary (traditional) law in the formal setting and it is established according to land law. Any willing agreements made and signed by two claimant parties and witnessed by the shemagelewoch shengo are considered as final decision.

Land administration committee members are volunteers elected for the implementation of the land administration system in ANRS. The participation of landholders was enabled through public hearings and meetings (direct participation) and through elected land administration committee members. The respondent committee members are satisfied with the support they get from the public authorities and from Kebele (parish) administration. The committee, on average, executes eleven cases per week. Committee members are happy with the support they give to local community and with the feedback they receive from the landholders. The committee member respondents reported that landholders are satisfied with the service from ANRS rural land administration system. The frequency of land related conflicts is declining since the introduction of land administration system in the ANRS.

Direct and indirect private sector involvement in ANRS rural land administration system is found to be minimal. The law depicted that private surveyors can be involved in cadastral surveying for the issuance of second level books of holdings. But in practice, both public and private sectors are not ready to assume the responsibility.

5.1.3 Operational Level

5.1.3.1 Land Tenure

Prior to the introduction of formal land administration system in ANRS, the landholders had no legal documents as evidence for their rights on the land they occupy. The neighboring landholders and people living in the area were the only evidences for the landholder claims of holding rights. During the formalization process the claims were approved by public hearings, where all landholders of the parish participated. The legal status about the land-landholder relationship was given after public approval. Between the public hearing and the public approval the party dissatisfied with the process has the right to appeal at the court. The adjudication process in ANRS is identified to be the combination of participatory and legal processes. It was possible to maintain both legality and legitimacy of land rights after formalization is completed. The legitimacy that emerged from participatory nature of the process helped to decrease the rate of conflicts.

The conflict rate generally decreased after the formalization process. However, latent conflicts come to surface at the initial stage of adjudication. Due to this some argued that the system is not successful in reducing conflicts. The current study identified that in ANRS the perception of occurrence of conflicts at the initial stage of implementation of the land administration system was much higher than the actual frequency.

The conflict rate in ANRS is still high. More than half (57%) of the individual farmer respondents reported that they experienced land related conflicts. The major reasons for conflicts, as reported by committee member respondents, were: inheritance related litigations (25%), boundary conflict (21%), rental (19%) contract related conflict (19%), easement related conflicts (11%), communal lands boundary conflicts (10%), and informal land sale related conflicts (9%) and plant shade related conflicts (5%). Conflicts related to informal land sale were reported to be difficult to solve due to the fact that land is legally not subject to sell. But according to the committee member respondents the problems related to illegal land sale are solved by shemaglewoch shengo using customary rules. The implementation of tenure core function is the key factor for conflict management.

Tenure core functions as the basic functions in rural land administration systems were rated higher than other functions. Important activities such as adjudication, transfer, lease and rental, updating, unique parcel identification, and establishing ground control points were rated higher. The good status of the rural land administration system of ANRS was reflected by the rating. A summary of the rating of the core functions by land administration professionals in ANRS is documented in *Table 4*.

Table 4: Summery rating for tenure core function (Number of total respondents=118)

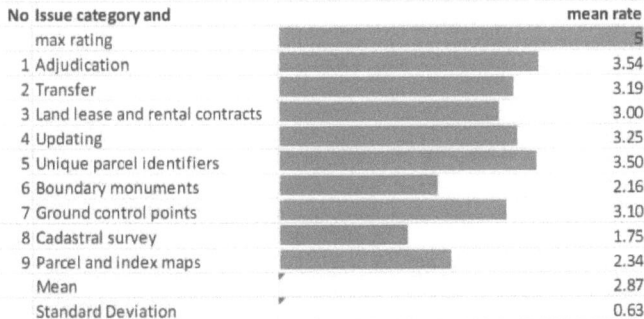

No	Issue category and max rating	mean rate
		5
1	Adjudication	3.54
2	Transfer	3.19
3	Land lease and rental contracts	3.00
4	Updating	3.25
5	Unique parcel identifiers	3.50
6	Boundary monuments	2.16
7	Ground control points	3.10
8	Cadastral survey	1.75
9	Parcel and index maps	2.34
	Mean	2.87
	Standard Deviation	0.63

The evaluation of land tenure of ANRS included also the registration system (both manual and computerized) and the cadaster. Five indicators for successful implementation of the ANRS land administration system were identified by BoEPLAU:

- Coverage increase in adjudication, registration and issuance of books of holdings;
- Protection for the weaker parties;
- Increased investment on natural resource conservation;
- Improved management on communal and state holdings; and
- Reduced land related conflicts;

Tenure insecurity in the region was related to land redistribution and is mainly the result of governments' interventions (Rahmato, 2005). The proportion of land holders reported to have tenure security by the study of Ethiopian Economic Association before the introduction of land administration system in ANRS was only 24% (EEA, 2002). This fits well to the estimations of Deininger et al. (2008) with outlined 27%. Contrary to this, all respondents reported that they have no fear of losing their holding rights. This finding is also confirmed by the high increase in the security feeling reported by the respondents. That needs to be interpreted in context: As said above the perception of tenure security is very much attached to land redistribution. The fact that there was no land redistribution in ANRS after the introduction of rural land administration system as well as several court rulings in favor of landholders with the book of holding increased the confidence of the respondents. However, as the causes for insecurity are changing over time and becoming condition dependent, the reported security felling cannot be considered as a final outcome.

The respondents also claim the steady increase in investment on natural resource conservation as evidence for the current high level of tenure security. But the achievements related to natural resource conservation were not free from externalities. Large scale campaigns for natural resources conservation are under implementation at the regional scale. It was a challenge in the study, to differentiate the contributions of the implementation of rural land administration system in the region from the effects of the campaigns.

The coverage increase in adjudication, registration and issuance of book of holdings is the main indicator for successful implementation of rural land administration system in ANRS. According to the most recent official report of BoEPLAU in less than ten years 3,624,424 holdings are registered and 3,132,879 books of holdings are issued in ANRS. Based on the report the regional coverage of the issuance of primary books of holdings was 98%. The average estimate of the regional coverage of the issuance of primary books of holdings by participants of the experts' panel in sample Woredas was 95%. The findings of the two sources can be said in agreement. The finding is also in agreement with previous comments by external evaluators about the fast speed, low cost and extended coverage of issuance of primary books of holdings in the ANRS (Deininger & Jin, 2006). It can be concluded that tenure security level increased with the introduction of the system. But it is too early now, to evaluate and to quantify the contribution of the system for sustainable land development in the region. Nevertheless, the study identified increased investment on natural resource conservation and proper land management.

The sample of individual farmers reported that they are satisfied with the land administration system of the ANRS. The respondents of individual farmers confirmed that the amount of tax they pay per year for their holding rights and updating and transfer costs are fair. The active participation of landholders was confirmed by the individual landholder respondents. But communal lands are not managed to the satisfaction of the individual landholder respondents (71%).

The protection of the weaker parties in ANRS rural land administration system is targeted at two levels (at system or design level and at operational level). ANRS rural land administration system attempts to anticipate major intervention areas for the protection of weaker parties and give legal protections at system level. Some practical problems were observed in the level of protection for weaker parties at operational level. The list of practical problems includes, but not limited to, forced illegal sell, limited information, prolonged litigation, etc.

Computerization of land administration records is one of the on-going activities in ANRS. The registration software called ISLA (Information System for Land Administration) was developed with financial support from SIDA (Swedish International Development Agency). ISLA is continually upgraded based on feedback from the field staff. ISLA is introduced in 74 Woredas and until now the data of 20 Woredas are fully digitized. For the details see *Table 5* below. Due to delayed legalization and limited coverage of computerized data in the region the expected efficiency gain is not fully realized. The delay in computerization can also negatively affect the implementation of cadastral surveys, as a spatial framework and the issuance of second level book of holding need computerized first registration data as input.

The major benefits of computerizing the first level registration data are facilitation for data cleaning and updating as a tool to identify parcels with ongoing conflicts and as a connection bridge between first level and second level registration data. The unique parcel identifiers were given during first stage of registration. The landholders in most cases do not know their parcel numbers. ISLA printout of first level data is a necessary tool for the cadastral surveyor to identify parcel numbers in the field and to facilitate the finding of names of landholder. Printouts also document the distribution of surveyed parcel and guide the planning and the control of field surveys. To summarize it, computerization is important for the speed, for the continuity and for the quality improvement of cadastral surveys.

Table 5: ISLA encoded data

#	Kind of Data	Total Number
1	Number of Landholders	1,372,065
2	Number of Parcels	6,038,914
3	Number of Weredas started	74
4	Number of Woredas completed	20
5	Number of Kebeles	1204

The spatial component is the basic part of cadastral systems. Lack of the spatial description (cadaster) was identified by the respondents as a major weakpoint of ANRS rural land administration system. The fact that a fully surveyed cadastral layer is too expensive at a particular stage in a country's development or in the development of part of a country, should not mean that documentation or registration of a diversity of rights over land cannot go ahead. The benefits and risks of when to start a cadaster need to be carefully weighed.

Currently cadaster in ANRS is limited to pilot projects and large scale irrigation project sites. According to the strategy document, second level certification and related cadastral land surveying are activities to be accomplished after careful preparation. Therefore the delay of establishing spatial framework is the result of ordering of activities based on available capacity and urgent needs. A network of 42 ground control points "called AM network" and linked with the national grid, is established. As the density and distribution of the ground control points in ANRS is not enough to realize the large scale second level certification program, the network of control points has to be extended and completed.

State land management is also one of the functions of land administration systems (UNECE, 1996). The size of state lands and the potential for investment is less in ANRS. The limited potential is due to high population density and increased coverage of small scale farms. Land banking is a method practiced by the ANRS to prepare land for investment and resettlement programs. In this program state holdings were demarcated and mapped. The mapped land is ready for transfer to investors (domestic and foreign) through lease contracts. Some national regional states such as Benshngul gumz and Gambela delegated the federal state to administer their state holding and attract foreign direct investment. In the case of ANRS no delegation was given to the federal state.

The state holdings management and investment areas are points of controversy that is termed as land grabbing by some (Rahmato, 2011). However the significance of state holding in ANRS is comparatively small, details of the distribution of investors on state lands in different districts are presented in *Figure 6* below. The problems reported by respondents are more related to capacity and follow up of contracts than to land grabbing itself. The maximum duration for the leasehold is 25 years. Lease contracts can be renewed after expiry period. In the lease contract the rights and obligations of the lessee as well as of the lessor are included. Among others, taking proper natural resources conservation measures are major responsibilities of the lessee. But in practice the large scale commercial investment farms are not environment friendly.

Figure 6: Share % of investment land area for Woreda, (source: BoEPLAUA 2010)

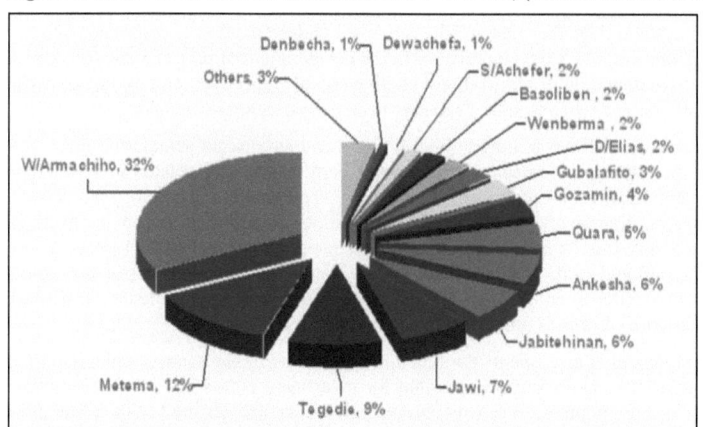

The strength of the system is the attempt to manage state holdings through clear legal provision and lease contracts. But due to capacity limitation, legal provisions and lease contracts are not fully enforced. The pace and quality of land preparation is, as reported by the regional investment promotion agency, not in balance with the demand for land investments. Therefore, large numbers of applications for land investments are pending at BoEPLAU.

5.1.3.2 Land Valuation and Expropriation

Valuation can be performed for compensation payment estimation, for taxation and for market (UNECE, 2001). Land valuation in ANRS is limited only to compensation payments for land expropriated to public purposes. The study identified that land valuation activities in the Amhara region are based on procedures stipulated in law. The expropriation of holdings is defined as '... *taking the rural land from the holder or user for the sake of public interest paying compensation in advance by the government bodies, private investors, cooperative societies, or other bodies to undertake development activities by the decision of the government body vested with power*' (ANRS, 2006).

The compensation has to be paid in advance of taking possession of the land. In practice there are lots of cases, where land holders are expropriated before compensation payments are due. On the contrary, there are instances, where compensation is paid to landholders, but the land is not taken for the development purpose due to delayed projects. Even if the legal right after paying compensation is by the project, the previous landholder is allowed to use the land until the project is realized. The expropriation of holdings resulted in landlessness. In ANRS the system and practice of rehabilitating the landless people is very weak (Ambaye, 2013).

The payable amount of compensation is calculated simply by multiplying the average income of the recent past five years with the factor 10. This method of valuating the land is a permanent source of controversy. The respondents of the survey also articulated their dissatisfaction with the amount of compensation payment for expropriated holdings. The literature review and survey results revealed that the main issue of expropriation is the quantum of compensation (Alemu, 2012; Ambaye, 2013). The fact that land valuation is limited only to land compensation activities hints all approaches to improve the method of land and property valuation.

The result of the experts' panel documents that the high of compensation mainly is based on the requirements given by federal compensation law. The experts reported that the paid amount is very low and cannot be considered as fair compensation. This fact also was confirmed in the interviews with the individual landholders. Uncertainty about the timing and amount of compensation is more damaging than anything else (Van Den Brink, 2002). See Table 6 below for the details.

Table 6: Summery table for the rating of value core function

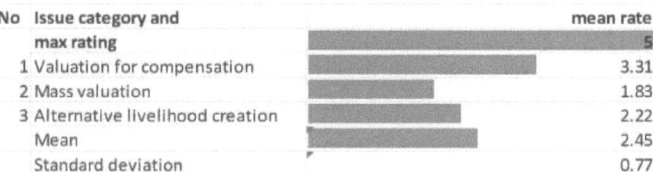

No	Issue category and max rating	mean rate
		5
1	Valuation for compensation	3.31
2	Mass valuation	1.83
3	Alternative livelihood creation	2.22
	Mean	2.45
	Standard deviation	0.77

Mass valuation is not practiced in ANRS and there is no valuation procedure for taxation. The taxation system for rural land is based on the potential productivity of soils, but data and methodologies to get figures on the productivity of holdings are missing (Chole, 1990). Nevertheless, the respondents of this study (landholders) consider the amount of tax they pay for their holding right as fair.

Mortgage according to land holdigs is a tool to make dead capital alive and it is the way to capitalism (De Soto, 2000). However the contribution of mortgage for economic growth in ANRS is minimal. Mortgages in the ANRS land law are limited to 'investors', who are leasing state holdings for a defined period from the government. The contribution of investors compared to large number of small scale farmers in ANRS is insignificant. The landholders normally lack the capacity to define convincing projects, which is a prerequisite to formal banks to approve loan. Additionally, the peasant farmers often live in a scattered way and the banks believe the administrative costs to manage loans from such locations to be very high. For these reasons and probably some more, the formal financial institutions are not ready or willing to address the issue of mortgage for small scale farmers. This is in conformity to the findings in literature (Deininger & Jin, 2006). Titling of land as a measure for increased mortgage is overrated in countries of the south. Financial institutions do not necessarily give applicants credit only because they possess title certificates (Obeng-Odoom, 2012). Banks take titles for much the same reason that kidnappers take hostages – titles mean a great deal to the party from whom payment is desired (the hostage giver), while having little value to the hostage taker (Kronman, 1985).

Credit and saving associations like ACSI (ANRS credit and saving institute), who are lending money by organizing group of people to control each without any other form of collateral, are trying to fill the capital need of the small scale farmers. ACSI is on a preparatory phase to lend money using the second level certificate as collateral. This can be seen as a great breakthrough for land transaction in ANRS. Currently land transaction in ANRS is confined to rental market.

The introduction of land administration system in ANRS contributed for the increased involvement of landholders in land rental market. The significance of rental market for the landholders is very high. Many of the individual farmer respondents (88%) are involved in rental land market. The entire rental contracts are officially registered. The rental income share for the landholders increased from 30% to 50% in the past ten years. The involvement of landholders in rental market before the introduction of land administration system in ANRS was only 5% (EEA, 2002).

5.1.3.3 Land Use Planning

Unlike the recommendations found in literature (Williamson et al., 2010) land use planning functions are weak-points in ANRS rural land administration system. According to the regional land law the implementation of approved land use plans on all holding types is mandatory. But no detail land use plans are prepared and approved by BoEPLAU so far. In the discussion the experts of BoEPLAU explained that the participatory land use plan will be prepared in two stages. The first draft of the land use plan can be derived from existing data acquired during the first registration. Based these data, the Kebele experts will make agreements of use at parcel level with landholders. After receiving their approval the participatory land use plan will get legal force. In a second stage geo-data – gained by the cadastral survey for upgrading the primary book of holdings – will be used as input to develop and improve the land use plans. The guideline to support the implementation of the planned activity is approved.

5.1.3.4 Development Control

Development control was not practiced on rural lands in ANRS. 'The lack of detail land use plans as regulatory tool is the main reason for the low achievement of this defined core function. The impact of having no development control mechanism in ANRS can be seen by wide spread of informal settlements especially on road sides and around kebele centers, encroachments as well as by misuse of communal holdings and Eucalyptus tree plantations on fertile agricultural fields. Respondents confirmed the miss use of individual holdings due to lack of development control tools.

In the case of the ANRS, all the respondents in the experts' panel reported that no significant activity is performed related to land use planning and land development control core functions. But the redistribution of hillside slopes to landless youth groups in Wollo and Gonder is considered to be a good beginning by the respondents.

5.1.4 External Factors

Personnel development as part of the general capacity building program was an essential contribution for the sustainability of the land administration system. In collaboration with SIDA and the Royal Institute of Technology (KTH) of Sweden, a total of 24 land administration professionals from BoEPLAU and from Bahir Dar University were trained at MSc level.

ILA (Institute of Land Administration) has intake capacity of forty students per year and in 2009 the first students graduated. Additionally, the Ethiopian Land Administration Professionals Association was established and hosted in ILA. The association is contributing for technical development of land administration system in the country.

The number of professionals assigned to undertake land administration responsibilities is increasing (see *Figure 9*). 87% of respondents stated that within the last ten years the budget assigned for land administration offices was increasing. Nevertheless, the current status of Woreda and Kebele office as well as equipment was rated as very poor by 80% of the respondents.

The respondents' assessment is in line with official reports. Human resource development at BoEPLAU is carried out in two ways. The first is in-house trainings and workshops with a duration of 5 to 10 days every year. The second one is a long term professional training. All trainings are focused on surveying, computerization, registration and general land administration concepts. According to 67% of the respondents the number of on-job and in-service trainings was increasing.

Figure 7: Human power and budget increase in sample Woredas (district level)

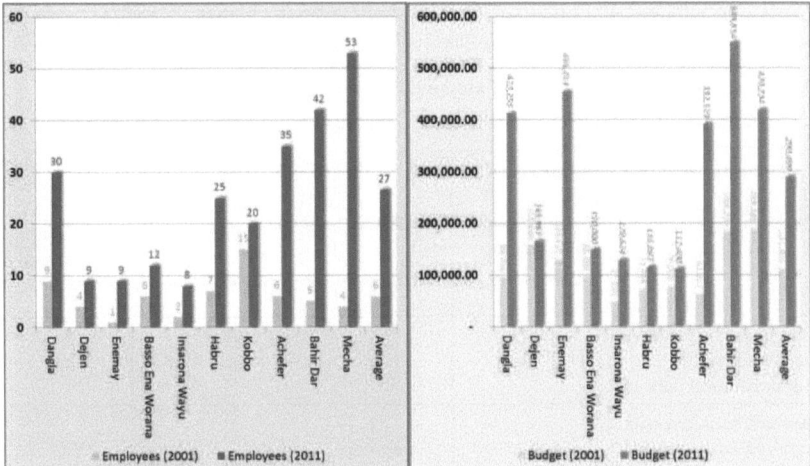

The main stakeholder sectors appreciated the introduction of land administration system in ANRS and they identified the following major effects strongly linked with the land administration system:
- Natural resource conservation and development structures are becoming sustainable;
- Legal research is facilitated as data are created in land offices;
- Delivery of land to investors is now based on information in the land data base;
- Land taxes are based on information in the land offices;
- Urban development and urban expansion is initiated based on data from land offices.

Within the current study stakeholder offices also were asked to articulate recorded changes in their institutions caused by the introduction of the land administration system in the region. The following changes are described as the most important ones:
- For decisions on land related conflicts, the justice system is now dependent on evidences from the land sector;
- The procedure to give license to investors is revised by the investment agency;
- For town expansion plans compensation payments become a must and a budgetary burden.

Contributions expected from the land administration system for efficiency and effectiveness of the stakeholders' responsibilities were (selection):
- Support through guidelines and strategies;
- Proper judgment through provision of land holding certificates;
- Land banking started;
- Clear and up-to-date information about landholders.

Representatives of major stakeholder offices reported significant positive changes as a result of the implementation of the land administration system in ANRS. However, the design and implementation of land administration system in ANRS was not supported by extended activities. The development was guided by a step by step approach to learn from experience. Continuous piloting and testing for experiential learning was the strength of the system. A summary of responses from the major stakeholders' institutions is documented in *Table 7*.

Table 7: Summary of the replies of major stakeholder institutions

Questions	Replies				
	Bureau of Agriculture	Bureau of Justice	Trade Industry and Investment Promotion	Revenue Authority	Bureau of Urban and Industry Development
Does your office have access to land administration data set?	No	No	No	Yes	Yes
Are you satisfied with services you get from the land sector?	No	Yes	No	Yes	Yes
Do you believe that land administration system has any contribution for accomplishments of your tasks?	Yes	Yes	Yes	Yes	Yes
Do land administration activities create any threat to any of your office interests?	No answer	No	No	No	Yes
What are the contributions of spatial data sets for accomplishments of the responsibilities of your organization?	to identify watershed boundary and forest cover and distribution map	to resolve boundary disputes	No answer	Not yet implemented	to integrate rural kebeles into town cadasters
Do you have any quality requirements for spatial data sets?	relevant and quality data	No answer	No answer	No answer	Yes
How much can your organization pay for land administration data sets	not willing to pay	No answer	No answer	No answer	No answer

5.1.5 Review Processes

The review process carried out by BoEPLAU stated the rural land administration system as strong. Every year after the completion of annual reports results of pilot projects were evaluated by external reviewers with the financial support from SIDA and USAID. In the last ten years three major institutional rearrangements have been made to address the changing needs in the sector. The last restructuring was made after the completion of a BPR (Business Process Reengineering) study. BSC (Balanced Score Card) is used as a tool to plan and assign responsibilities. The allocation of Kebele level land administration experts is one of the consequences of BPR and BSC studies.

BoEPLAU of ANRS developed a strategy document in 2003 and made amendments in 2007. The development of the first strategy document was based on the lessons from pilot projects. The strategy was continuously. Customer satisfaction surveys were not regularly outlined by BoEPLAU. But in the current study the system users testify a high level of satisfaction with the land administration system.

The SWOT analysis of ANRS rural land administration system was carried out during the experts' panels. The strengths and weaknesses of the system can be classified in cadastral surveying related, protection of weaker party's related, organizational setting related and land valuation related topics.

The main strengths related to cadastral surveying and land registration are:
- Participatory approach during adjudication process;
- Regular updating of land records;
- The mechanism in place for the exchange of data with revenue collection authority for tax collection (it helped to have a wider tax collection base and fair taxation);
- The introduction of a computerized system at early stage of system development;
- The capacity of the system to generate multipurpose data;
- The contribution for increased long term investment and environmental protection;
- Strong public support and customer satisfaction;
- The role for increased public awareness on land administration issues;
- Low cost and rapid registration system as a step by step processes;
- The inbuilt mechanisms to experiential learning.

Strengths of the system related to land valuation core function were also compiled. As main strengths the following were specified:
- The valuation of land is strictly regulated by federal law and regional directive;
- Public consultation and general agreement is required by law before any form of expropriation (this rule is not strictly followed in practice);
- Expropriation is permitted only if the planned activity is beneficial for the public; However the definition of public advantage is very wide and debatable;
- Preparation of development projects for affected landholders.

But there are also some limitations to the system:
- Excessive state power on land used causes tenure insecurity.
- Expropriation of holdings without any form of compensation in the name of public benefits is the most eroding force of tenure security;

- Though landholders are entitled to get compensation before land is taken away from them, delayed or no compensation under the cover of public purposes and budget shortage are still posing problems for the landholders.

Organizational setting is one of the key factors that can make or break a system. All in one type of institutions are recommended as a model organizational arrangement for land administration systems. The type of organization for land administration is influenced by historical factors and whether the back ground of the country legal system is from common law or civil law. The role and capacity of notaries and insurance organizations also can affect the type of land administration institution. In the case of ANRS, the role and functions of insurance companies and notaries with regard to rural land administration is negligible. The main strengths of BoEPLAU as land administration institution as stated by respondents are:

- Official structure is established down to the grass root level (Parish level);
- The institutional setting has a strong linkage to land administration and land use committees as well as to shemagelewoch shengo (the representatives of the landholders);
- Core land administration functions are under the responsibility of a single institution, namely BoEPLAU;
- The feedback and control system is well established in the institution;
- BoEPLAU is represented in the regional cabinet, which is the highest governmental decision making body headed by the regional president;
- The system recognizes traditional rules through shemaglewoch shengo (arbitration committee);
- Prime movers of the system are volunteer committees at kebele and sub kebele level.

The land administration system of ANRS also was evaluated based with regards to its ability to give legal protection for the weaker parties. The weaker parties in the society include the elderly, children, physically disabled people and women. Land related conflicts and competition to benefit from land resources is very high in ANRS. Strong legal mechanism and practice is needed to protect the equitable share of the weaker parties in the society. The respondents identified the following strengths of ANRS land administration system related to the protection of the weaker parties:

- The availability of clearly defined articles in the land law to protect the weaker parties;
- Women land holding right is not against the tradition;
- The good culture of protecting and respecting the elderly people;
- Photos of both spouse attached on the book of holding to legally recognize the equal rights of both;
- The participation of women in land administration committees.

The major weaknesses of the land administration system in ANRS identified by the respondents were also categorized into the same topic groups. The main findings are documented below:

Related to cadastral surveying and land registration:

- Land administration activity in ANRS is not yet supported by spatial data;
- The lack of historical documents to be used as a reference during adjudication and conflict;
- The experience of the staff on land registration and cadastral surveys was minimal as the subject is a new exercise in ANRS;

- The participatory way of working developed was not strictly followed especially at the beginning;
- Updating rate of the land law was slow.

Land valuation:
- The valuation was done only for compensation purposes;
- The amount of compensation that can be calculated based on the provisions in the law is not fair for land holders.

Organizational setting:
- Lack of proper professional to fill the job positions in the organogram;
- Equipment and supplies are limited;
- Campaign approach is overused;
- The organizational structure of the institution was not based on the core functions of land administration.

Protection for weaker parties:
- Participation is still limited
- The traditional farming system is demanding and as a result of this weaker parties are forced to rent their holdings

The major opportunities of land administration system in ANRS reported by the respondents were summarized as follows.

Opportunities related to political and legal environment:
- Laws to govern land administration are in place and the political leadership is committed to implement the policy;
- Land is identified by the leadership as one of the country's key resources for development, this leads to proper attention to the sector;
- Land administration is part of the 5 year development and transformation plan;
- Significant attention is given in 5 years development and transformation plan for land administration specifically to second level certification program;
- Major land issues are enshrined in the regional and federal constitutions;
- The presence of stable state structure to implement land administration in ANRS.

Opportunities for land administration system in ANRS related to the economic environment:
- Government's capacity to implement land administration system increased;
- Higher level of transactions demand good land administration data,
- Landholders' capacity to pay for land administration services strengthened;
- Additional infrastructure and services in place,
- Land holders' capacity to invest on land increased;
- Increased investment requires efficient land administration system,
- Fast growing economy needs to have progressive land administration system in place;
- Budget allocated for land sector increased:

Social related opportunities to land administration system in ANRS:
- Decreased conflicts increased social bondage;
- Public hearings and meetings increased socialization,

- The productivity of communal holdings increased ;
- The bond between social norms and land administration organization is strengthened;
- Capacity of land administration system to incorporate traditional rules increased.
- Public support to the sector increased

Opportunities related to technological development identified by respondents:
- New computerized system (Information system for land administration – ISLA) developed;
- Modern survey equipment introduced;
- Access to modern communication facilities increased;
- Cost of survey equipment is decreasing;
- User friendly equipment and systems are developed.

The major threats of the system discussed in the panels were:

Political and legal environment related threats:
- Framework laws are not updated to reflect regional interests;
- Low level of attention to land administration activities by zone and woreda administrators;
- Lead institution at the federal level is not strong;
- Different system for rural and urban lands;
- Over-ambitious plans not considering preparatory steps on time.

Economic related threats:
- Increased number of conflicts concerning small land with an increased value of land;
- Increased number of absentee landholders;
- Number of expropriation affected landholders increased;
- Land grabbing;
- Foreign direct investments on the expense of indigenous people and natural resources;
- Increased demand for urban space creates conflict with rural use.

Threats with social significance:
- Weakened local rules and weakened indigenous organizations;
- Increased individualism;
- Limited social controls:
- Urban sprawl at the expense of rural community;
- Pastoralists rights are very difficult to adjudicate;
- Not properly managed settlement structure of indigenous people and shifting cultivation.

Technology related threats:
- Technological advancement demands frequent instrument upgrading;
- Programs and methods become obsolete very fast;
- Increased requirement for trained professionals;
- LAS becomes dependent on other agencies;
- Data security become difficult

The summery of the result of the SWOT analysis based on the input from the respondents in the expert panels is presented in *Table 8*.

Table 8: Summery of SWOT analysis

Strengths	Weaknesses

• Land policy is of constitutional category • System responds to the needs of society= participatory and responsive system • System is equitable for all • System is economically viable - low cost methods were applied • All in one type organizational structure • BoEPLAU is legally established with clearly defined tasks • Strategies that are appropriate to reach and satisfy objectives - Book of holdings with two distinct strata • Mandates are given by law and well communicated with users • Participatory, fast and low cost systematic registration • Continued human resource development program. • Focus on capacity building. • Professional association is established • Ongoing trials and pilots • The system was regularly evaluated	• Experts at all levels have little background knowledge about mission, vision and strategy • Private sector involvement is minimal or no • Organizational structure is not based on land administration core processes and main functions • Valuation is only for compensation purposes • Taxation is not based on land valuation • No land use planning and planning control • No systematic and regular customer satisfaction surveys were conducted
Opportunities	**Threats**
• Relative peace and security all over the regional state • Decentralized government structure and decision making • Stronger commitment to good governance and rule of law • Political commitment • Well-functioning land rental market • Incentives and support for long term investment • Strong and well-functioning cultural conflict management culture • Rapid infrastructure and technological development • Very strong users support	• Influences towards to blanket standardization by the federal authorities and donors • Weak control over the grass root level leaders • Disadvantaged position in globalization • Enhanced natural resources degradation • Land grabbing by investors and unplanned settlement programs • Global economic crisis • Dependency on imported technologies • Harmful traditional practices • Uncontrolled population growth

After discussing strengths, weaknesses, opportunities, and threats of land administration system in ANRS the respondents proposed major changes on the system. The major changes proposed by the respondents are:
- Legal experts have to be assigned at Woredas level;

- Computerization of data should be launched from the Kebele level;
- Land use planning and land use control procedures have to be developed;
- Strong person-power development strategy has to be crafted;
- Increased budget allocation;
- Increased in-service training;
- Increased supply of transportation facilities and equipment;
- Remuneration and proper incentive mechanisms for the staff;
- Increased supervision and technical support.

The significant consideration given to major land issues in the constitution is an advantage at the policy level. This can be a lesson for other African countries while developing a new land administration system. The institutional mandate should be based on core land administration functions. Institutions need to consider customer satisfaction surveys as indicators for achievements.

Professionals and managers from land administration offices of 118 different Woredas rated the status of specific tasks of ANRS rural land administration system. The summery results of this questionnaire-based survey are presented in *Table 9*. The detailed questionnaire is documented in the appendix.

Table 9: Ratings given by land administration professionals on the status of different land administration activities in the ANRS (5 good/sufficient - 1 weak / insufficient)

No	Activity	Summery rating in %					Average rating
		1	2	3	4	5	
1	**Cross Cutting Issues**						
1.1	Policy formulation & monitoring	12.9	4.0	11.8	16.1	55.2	**4.0**
1.2	Legal	15.1	12.2	22.6	15.4	34.7	**3.4**
1.3	Computerization and IT Systems	56.4	13.6	12.6	8.6	8.8	**2.0**
1.4	Research & Development	71.9	10.1	11.5	3.6	2.9	**1.6**
1.5	Capacity building	48.6	19.6	18.8	7.6	5.4	**2.0**
1.6	Public information and awareness	34.2	16.3	26.9	16.3	6.3	**2.4**
1.7	Finance and cost recovery	38.0	12.4	21.3	10.4	17.9	**2.6**
1.8	Gender issues	9.6	11.0	18.2	22.5	38.7	**3.7**
2	**Land Tenure**						
2.1	Adjudication	14.1	10.5	18.9	17.0	39.5	**3.6**
2.2	Transfer	19.6	11.2	22.5	19.0	27.7	**3.2**
2.3	Land lease and rental contracts	28.6	9.4	17.4	14.5	30.1	**3.1**
2.4	Updating	15.1	8.3	28.1	30.2	18.3	**3.3**
2.5	Unique parcel identifiers	22.4	2.9	9.5	13.3	51.9	**3.7**
2.6	Boundary monuments	44.8	18.1	19.5	8.6	9.0	**2.2**
2.7	Ground control points	23.5	7.8	17.6	22.1	29.0	**3.3**
2.8	Cadastral survey	65.9	7.1	8.2	9.1	9.7	**1.9**
2.9	Parcel and index maps	49.1	8.1	9.6	10.8	22.4	**2.5**

3	Land Value						
3.1	Valuation for compensation	23.6	6.7	18.4	16.9	34.4	**3.3**
3.2	Mass valuation	62.9	10.7	11.4	5.0	10.0	**1.9**
3.3	Alternative livelihood creation	47.9	13.4	15.3	8.6	14.8	**2.3**
4	**Land Use**	46.1	3.7	12.1	8.1	30.0	**2.7**
5	**Development control**	47.8	10.9	14.5	11.6	15.2	**2.4**

The study identifies, that large scale implementation of land use plans is constrained by technical complications and costs. It is believed that – in a first stage - land use plans can be prepared at parcel level by contractual agreement with the responsible landholders. After the land surveying is completed, local level participatory land use plans can be upgraded to full-fledged detail plans by linking it with cadastral maps. Attempts to make decisions on development activities without any data input from land use plans can lead to subjective and unfair decisions. In the worst case it can be a threat for good land governance and can erode the public trust on land administration institutions.

5.2 Description and Comparison of the Formal and Informal Property Right Systems

Proper description and understanding of the whole property system and its history is necessary before proposing improvements to the system.

The land administration system in the ANRS is strongly affected by both formal and informal property right systems.

5.2.1 The Hierarchy of the Legal System in Ethiopia

The highest governing law in Ethiopia is the constitution. The federal government and the member states have the legislative, executive, and judicial power. The house of people's representatives is the highest power, to be elected every five years by direct popular vote. There is also similarly organized law making body at the regional state level commonly known as the state council (FDRE, 1995).

The hierarchy of the legal system is generally organized in two levels, namely the federal state laws and the regional state laws. The formal land administration system is part of this hierarchy of laws. The member states have the right to develop their own constitution. The major objective of developing a constitution at the regional state level is to have a possibility for modifications based on site specific situations and peculiar needs (FDRE, 1995). The most pronouncedly used legal institutions in Ethiopia are the court, the civil administration and the local organizations.

The regional laws have to comply with the federal framework laws. In cases of contradiction between the two legal strata, the federal level has always the overriding power (Andersson, 2005; Ambaye, 2013).

The cabinet of ministers at federal level – under the leadership of the prime minister– and the regional cabinet – composed of selected bureau heads chaired by the regional presidents– are the two major executive bodies responsible for enacting regulations.

Directives are other important components of the legal system. The directives shall be developed based on the regulation. The responsibility for developing directives is mostly given to the implementing ministerial office at the federal level or to the bureau or to authorities at the regional level. The directives are supposed to show the exact hand and arm movement in the implementation of the higher level laws. The formal property right system in Ethiopia is governed by laws at all level of the hierarchy (see *Figure 8*).

In general, the legal system of each nation is based on the socio economic and political setting of the country. Land laws are very much under the influence of the local situation. It is nearly impossible to copy the legal system of another country as it is. But the philosophy and the rationale behind each article can be shared as a lesson for the development of site specific land laws.

Figure 8: Hierarchy of federal and regional laws in Ethiopia (Source: Anderson, 2005)

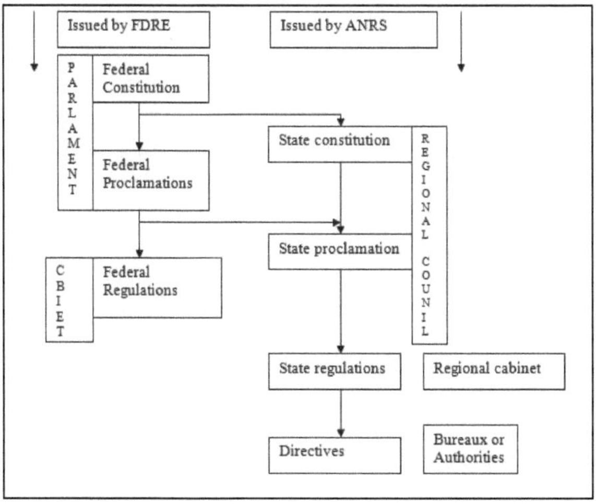

Acronyms:

ANRS-Amhara National Regional State

FDRE-Federal Democratic Republic of Ethiopia

CBIET- Cabinet

5.2.2 Formal Landholding Right

In ANRS, the formal land administration system is designed and implemented on basis of the federal and regional laws, namely the federal land proclamation (456/2005) and the regional land law (133/2006).

The constitution boldly underlines that the right to sell or buy land is not included in the bundle of rights given to the landholder. However, the landholder can be the owner of both movable and immovable properties developed on his land. The intention of legislators, while restricting the ownership right, is to protect the peasants from eviction caused by distress sell (as proved by the results of the questionnaire and in discussion with experts).

The landholders shall never lose the occupation without proper compensation, though some variability is reported in practice (Yersaw, 2012; Ambaye, 2013). The constitution states the subject in point as *'Ethiopian peasants have the right to obtain land without payment and the protection against eviction from their possession...'* (FDRE, 1995).

The right to hold property is also stated in the constitution Article 40/7. This right is given for every Ethiopian and the protection includes all immovable improvements made by the citizen. The improvements can be caused by citizen's labor, creativity or capital inputs on land. The rights of the citizen include the right to alienate, to bequeath, to transfer, and to remove his property, when the right to use the land expires (FDRE, 1995).

The federal framework land law, based on the constitution, defines holding right as the right of any peasant farmer, semi-pastoralist, or pastoralist, to use the rural land for the purpose of agriculture and natural resource development. It allows to lease and to bequeath the land to members of his family or other lawful heirs. It includes the right to acquire property produced on his land by his labor or capital and to sell, exchange and bequeath same (FDRE, 2005). In Ethiopian context, holding right refers to the right given on land. Property produced on this land refers to fixtures. Fixtures – contrary to the holding right – are subjects to sell. Due to this vague statement, landholders, especially in the urban areas, are capable of transferring their right on land together with a building or a house, or any improvement on land during sell.

Most property related laws, including land laws, are very much influenced by the civil code of the country promulgated in 1960. The socio-political setting in the country changed very much since the enactment of the civil code. As a result, some of the provisions in the civil code are outdated and not applicable. However, the definition of immovable in the civil code is still valid. The civil code defines the immovable as lands and buildings. Fixtures in the civil code are termed as intrinsic elements of goods. These elements include anything that by custom is believed to be a part of a thing and things that are materially united with a thing. Trees and crops are also intrinsic elements of a thing (EoE, 1960).

The holding right is the highest right for the holder that encompasses all transfer rights except land sale. The right normally has no time limit and hence it is different from a lease system. It is different from the use right, as the use right can be obtained by renting land from the landholders or the state. The use right is for an agreed and defined period. The maximum period is 25 years in the case of ANRS, but – as reported in the expert panels – the terms can be extended by the agreement between the two involved parties.

The holding right of any person is respected by law. No person shall be expropriated unless it is done by re-distribution according to the decision of people or for the purpose of public interest. The term public interest is often debatable. Adequate compensation is supposed to be paid for expropriated land before land acquisition. The controversy is on what is adequate for the subsistence farmer, whose life is entirely dependent on his holding (Yersaw, 2012; Ambaye, 2013).

Formal property rights are those that are explicitly acknowledged by the state and which may need government authorities for enforcement (Williamson et al., 2010; Dale & McLaughlin, 1999; FAO, 2002).

A better understanding of rights and restrictions linked to holding rights requires a detailed model addressing benefits and limitations on the holding right as well as the legal origins of rights and obligations. The model has to cover all aspects like beneficial rights, limiting obligations, public advantages and public regulations. The existence of the common right, the right on others' property, the right on users, the latent right, and the collateral right as a benefit or as an obligation has to be considered, as well as the fact that rules and regulations can cover the issue completely or partially.

In the current thesis the Legal Cadastral Domain Model (LCDM) of Paasch (2012) was adapted to the situation of ANRS. *Figure 9* shows the result gained in an analysis of the Ethiopian and ANRS law and considers the findings of the interviews with experts and farmers.

The beneficial rights of the holding right are the common right, the right on others' property, the right on users, the latent right, and the collateral right. The term limiting rights used by Paasch (2012) was changed to obligations in accordance with a proper description of the relation in ANRS rural land administration system. The types of obligations on the holding right are common rights, the right on others property, the right on users, the latent right, and the collateral right. Brief definitions for both, beneficial rights and obligations, are given below. The definitions are tailored to the situation in the ANRS.

Common right is the right to use a parcel in common. The use can be issued for short time, e.g. free grazing on crop lands, or for unlimited time, e.g. common pastures, community forests, and service areas. The right to use commons is related to the membership to the local community.

Right on others' property is the use right of the dominant land holder on the holding of the servant land holder. Examples of rights on others' property are right of way and easements. The servant holding is usually compensated for possible losses.

The right on users is the right of the land holders on the users of their holdings. It is the obligations of the tenants to serve the land lord. The service can be carried out in form of labor, extra holiday gifts, or of material supply for different occasions. This type of relation was totally abolished during the Derg era.

Figure 9: Modified legal cadastral domain model (LCDM) representing ANRS formal legal system (based on Paasch, 2012)

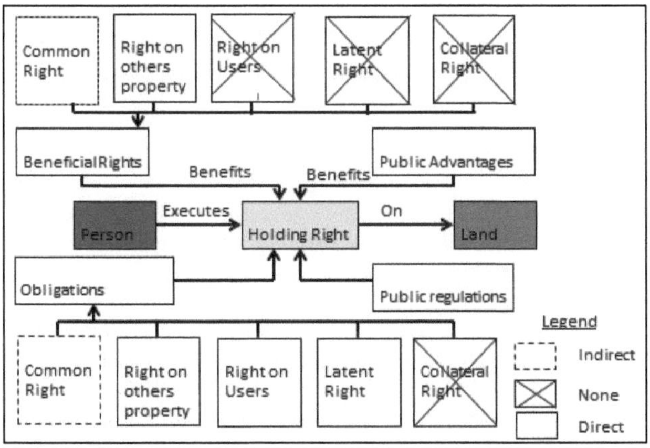

Latent right is the *'right not yet executed on a real property'* (Paasch 2012).

Collateral right is the right to borrow money from financial institutions or individuals by using the holding right as a guarantee.

There is no land in ANRS without any designated holder. The holder of the land can be a natural person, a legal person, a group of people, or the state. According to the proclamation 133/2006, landholder is defined as *'an individual, group of people or community, government body, social institution, or other body with a legal personality having a holding right over rural land'* (ANRS, 2006). As confirmed by experts, key informants, and individual farmers open access areas in remote locations de facto belong to the state lands.

The holding right is linked with beneficial rights and obligations, which may differ between the different kinds of landholders. So, e.g., the state holdings can be transferred to investors by lease contracts. The investors can be domestic or foreign individuals or companies with an investment license for doing business in Ethiopia. The transfer of holdings under service, given to institutions like schools, hospitals etc. are limited to landholders, who can legally run the outlined service.

Experts reported that obligations imposed on the holding right are more pronounced than obligations on ownership right. This is caused by the fact that land sale is not allowed in the holding right.

Proclamation 133/2006 defines the common holding as rural land not under the ownership of the government or of any private holding, but used by the local people in common for grazing, forestry and other social services. In most cases, communal holdings are governed by traditional rules and by-laws. As stated by experts and farmers, the traditional administrative mechanisms are acknowledged by the land law of the region to reduce conflicts caused by resource competition. According to the regulation the local society is entitled to establish by-laws considering the local circumstances. The decisions based on these local rules are legally valid unless they are not in contradiction with established formal law.

The regional land law has also regulations, how to transform communal holdings into individual holdings. Legal restrictions in the transformation process are the agreement of the legal users of the concerned area and the perpetuation of the existing land use type after individualization. Additionally, the transformation process has to be approved by the authorities to minimize possible environmental consequences (ANRS, 2006).

The common rights in ANRS land law are connected with a Kebele (parish) membership, but in some cases the rights are limited to specific groups within the Kebele. Also the landless dwellers of the Kebele have full right and responsibility to use the common pool resources within their vicinity.

Common pool resources in the region are grazing lands, community and conservation forests, market places and other service areas, river banks, and water bodies. In practice – as told by the interviewed experts – the right to use the common pool resource is not exclusively given to the landholding rights in ANRS legal system, but it is a beneficial right with a weak connection to the holding right.

Land owners frequently desire to restrict the access of others to their holdings, often caused by the need for privacy and territorial imperatives. The local society and the state attempts to regulate these needs for the benefit of other community members. The relationship can be termed as the property owners' golden rule, saying: *'I shall use my property as I think fit. The authorities must not interfere in my activities. My neighbors may use their properties, as long as they don't cause me any harm. It is the authorities' duty to protect me from my neighbors.'* (Kalbro, 1996).

However, regulations of the ANRS land law allow the holder to use other holders' land, if he has no other possibility to access the public infrastructure, e.g. roads. Another case to use others' land is to pass runoff of water, if the contour and drainage pattern demand it. And finally, a landholder has the right to establish irrigation channels on the land of his neighbors for watering the plot.

The rights of the holder described in the regulations are similar to the right of way and easements in other countries. The difference lies in the development of specific societies and in the variation of needs on infrastructure and services.

There is no provision or practice in ANRS land law, where a land holder has any right on the users of his property. He only has the right to rent his land. The maximum legally permitted period for a rental agreement between the landholder and the tenant in a single term is 25 years. The right for sublease of land is dependent on the prior contract agreement. The landholder has to be notified, if the tenant is subleasing the rented holding with the possibility to cancel the contract, if the land holder is against the subleasing.

The proclamation of other lower level laws, such as regulations, include no rights and/or obligations of the lessee, when the holding right is subleased. Therefore, conflicts related to subleasing have to be resolved by the provisions in the civil law. A clearly stated tenant protection provision is missing in the current ANRS legal system, as experts stated during the panel discussions.

In the legal system of ANRS, the holding right cannot entertain pre-emption rights due to the legal prohibition of land sale. Expropriation of the holding rights is possible, if land is needed for public services. The expropriation of holdings is defined as *'... taking the rural land from the holder or user for the sake of public interest paying compensation in advance by the government bodies, private investors, cooperative societies, or other bodies to undertake development activities by the decision of the government body vested with power'* (Ambaye, 2013; FDRE, 2005; Yersaw, 2012).

According to the law, the compensation has to be paid in advance. In practice – as reported by the experts – some of the projects causing the expropriation are delayed and the previous holders are using the plots until the projects are launched. This practice cannot be classified as a latent right as it is not legally permitted.

In ANRS mortgage law is limited to investors/land users, who are leasing rural land for a specific period from the government. Individual holding rights are given mostly to peasant farmers. Normally they are not able to outline convincing projects, which are a prerequisite for receiving a loan from the banks. Financial institutions – except micro finance institutions – are not willing to address the issue of mortgage for small scale farmers. Both, the financial institutions and the small scale farmers are not ready to practice mortgage. Therefore, at this point in time, it is practically meaningless to proclaim collateral rights, which are far away from being applied. However, capacity development of both small scale farmers and financial institutions is necessary to benefit from the contributions of mortgage for increased investment.

Public advantage and public regulation are interrelated. Normally public regulations are intended to protect the general benefits of the society and to contribute cohesion in the society. The public advantage to specific land holders accrues, if the public regulation is linked to a certain group of land holders. For example, the land use plan may restrict the upstream land holders to cultivate their land only with trees, perennials, or grass to protect the downstream users from damage caused by excess runoff. Another example in the rural context is the construction of diversion ditches. The ditch will occupy land from certain landholders for the benefit of many others. The public regulation imposed on certain holdings benefits many holdings.

The public advantage in ANRS often is obtained by the development of the area. Titled land makes a significant difference on the value of the holding compared to untitled land. Planning and infrastructure developments in areas have significant positive impact to land values and to rental amounts. As the benefit is not distributed in the same manner to all landholders in the area, a legal regulation for balancing this unfairness was proposed during the expert panels.

Most public regulations in the ANRS legal system are related to land use controls and environmental protection measures. These regulations are common to all concerns and are included in the land law. Landholders are obliged to implement the land use plans as developed for their area. According to the land law of ANRS, the local level participatory land use plans are binding after approval by authorities. The obligations of the landholder specified in the land law are targeting proper land use, sustainable development, and the protection of shorelines and riverbanks, e.g.:

- to plant trees at the boundaries of his holding;
- to control erosion using different technical mechanisms;
- to protect water sources and wet lands from drying out;
- to exercise proper care for wildlife and birds sheltered on his holding (ANRS, 2006), and
- to plough the land far from river or gully.

The public regulations in ANRS legal system have the power to benefit or to restrict the landholders. Therefore, public advantages as well as public regulation are identified in the Amhara formal system.

5.2.3 Informal landholding Right

One objective of this thesis was also to modify the LCDM developed by (Paasch, 2005; Paasch, 2011; Paasch, 2012) according to the existing rights and obligations of the informal system in ANRS.

In the traditional setting the community defines what activities are permitted and what not. Restrictions in the informal setting are imposed by the local society and by its culture rather than by the land law (Lemmen et al., 2009). It is not fencing or guarding a property that is important to assure ownership in the informal setup. The local society has to approve and to accept the act (FAO, 2002). The informal system employs a shared control, while a formal system relay on external forces to enforce decisions. Disregarding the informal rights of the local society by rating them as irrational relics of an early age is no more logical (Lane, 2001; Onoma, 2008; Obeng-Odoom, 2012).

The central right for the explanation of the informal setup of land issues in ANRS is the informal holding right. The relationship or the informal setup is – same as in the formal setup – the relation between the subject (person), holding right, and the object (land). The holding right in the informal setting allows the transfer of properties within the Irist holding group. It also enables land transfers outside the group, but only with the consent of the group leader. In contradiction to the formal law, the holding right of the informal law includes land sale and mortgage (Rahmato, 2005; Mesfin, 1991).

The major holding parties in the informal setting are individuals, groups of the community, the Orthodox Church, and other service giving institutions, such as the Kebele administration. The land holding parties are similar to those of the formal system. Most of the holdings under individual holdings are crop lands. The holdings of certain groups of the society are either forests or grazing lands.

The discussion on commons is about the benefits and burdens of individual landholders in relation to common pool resources management. The common pool resources are mostly linked with membership to the local society. The origin of the common pool resources is related to state lands of the imperial period. These sites were allocated by the state for different services as a compensation or payment. Until the fall of the Imperial regime this land was named as state land and – according to the legislation at that time – it was de facto an open access area in every Kebele. Sometimes the right to administer vacant spaces was given to the church in the Kebele.

According to the witnesses of the key informants of the current study, land plots without clearly defined owners were more rare in densely populated areas, e.g. in North Sheoa. During the Derg regime – after the fall of the Imperial system, Kebeles were established as a grassroots level of formal administration. Land holdings with no clear claimants, state lands, and/or lands allocated for the church were transformed into communal lands. In the informal setting of some regions, such as South Gonder or North Sheoa, land users still give the sell of certain products or pay some money for the communal land, which was allocated to the church during the Imperial period.

The current legislation in ANRS acknowledged the importance of the traditional rules for managing common pool resources. Investigations gave evidence of differences in the management of these resources dependent on the location. The differences are inherent to the tradition of local society and historical reasons. The rules in the informal setting are different, because the right to develop by-laws for administration of common pool resources was given to the local society. According to the proclamation 133/2006, by-laws governing common pool resources of a given local society can be different from others as long as they are not in contradiction to the formal law (ANRS, 2006). In some sites, such as North Sheoa or South Gondar, religious rules are influencing the management of common pool resources.

The common right in the formal setting can be said to be the reflection of the common right in the informal setting. The description and use of common rights in the informal setting are nearly identical to the common right described in the formal setting. But common right in the informal setting normally is not directly associated with the holding right of individuals.

According to the results of the group discussions, the right to freely graze animals on the holdings of others is a common practice all over the regional state. But the traditional rules for free grazing are varying between areas, as outlined investigations in the four sample areas brought to evidence. So, e.g., the starting date of the free grazing is different in all sample sites. The date is related to the agro ecological zones and the types of potential crops in the area. The definition of the date is based on the final crop harvest date and even this can vary depending on the length of the rainy season. The local societies have no clearly defined forum or delegated group to decide the date for the beginning of the free grazing. In practice, the agreement is reached on consensus every year and no conflict is reported on the seemingly vague decision making process. The landholders are obliged to collect their crop before the commencement of the free grazing. The landholders have the priority to graze their animals or to collect and to store the crop residue – if necessary – only until the beginning of the free grazing day. The free grazing right is an obligation on individual holdings as a servant. Free grazing right is also a benefit to graze on others' fields.

The free grazing right is not a localized right and it is not given to specific landholders. Every member of the local society can make a claim on the right. In this regard there is a similarity between common right and the free grazing right apart from the fact that the free grazing right is just for a defined period of time. This indicates that commons exist in the informal system as a beneficial right as well as an obligation (see *Figure 10*)

Figure 10: Legal Cadastral Domain Model (LCDM) representing the informal setting in ANRS (based on Paasch, 2012).

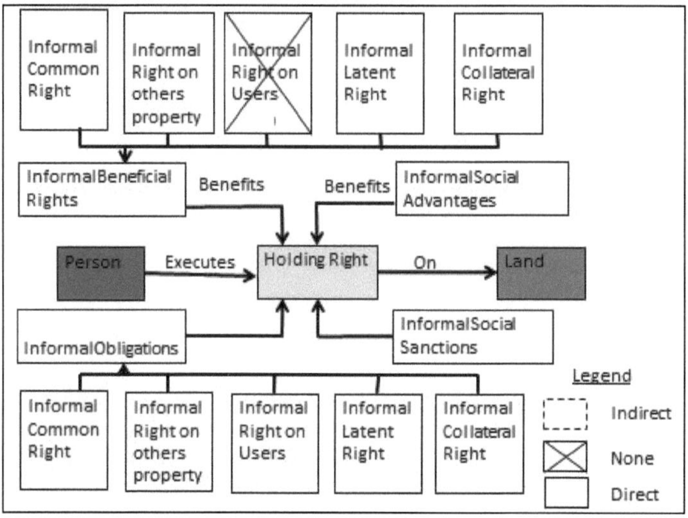

| Informal Common Right | Informal Right on others property | Informal Right on Users | Informal Latent Right | Informal Collateral Right |

| Informal Beneficial Rights | Benefits | | Benefits | Informal Social Advantages |

Person → Executes → Holding Right → On → Land

| Informal Obligations | | | | Informal Social Sanctions |

| Informal Common Right | Informal Right on others property | Informal Right on Users | Informal Latent Right | Informal Collateral Right |

Legend:
- Indirect (dashed)
- None (X)
- Direct (solid)

An informal right to use other's holdings exists as an obligation as well as a beneficial right. The relationship is direct. The right for the access to grazing lands, to water points, and to the main road, can be mentioned as examples for the informal right to use other's holdings.

The right on land users was abolished by the Derg proclamation (PMGE, 1975). During the Imperial period landlords had tenants (serfs) on their land. Landlords had the right to transfer their tenants together with the land or to order them to perform labor works on other locations. The landlords have the right to force the tenants to pay items that were not commonly included in the traditional agreements. These types of exploitative rights and relations were legally abolished during the Derg regime and they were also abolished in the informal setting. Personal right does not exist as a beneficial right in the informal setting. Nevertheless, personal right as obligation still exists – due to social protection of the rights of share croppers and of other informal land rentals.

Land sales during the Irist system was strictly regulated by the local society. The priority for sale was given to people of the same group of right holders. Outsiders only could buy land, if they were accepted by the group. The system therefore gave the pre-emption right to members of the group and accepted outsiders. The practice was abolished by the Derg proclamation.

Informal land sale is reported from all of the sample sites. According to the tradition, the previous landholder has an informal pre-emption right – before the most recent land redistribution. In some areas this right was extended to the former Irist right holders. In the informal system the current landholder is allowed for socially accepted sale by informing the previous landholders and the neighbors. He is allowed to sell the land to others, if the previous landholders and the neighbors are either not interested to buy the land or they are not able to pay the requested amount. By tradition, the current landholders are expected to accept a reasonable reduction of the price for the previous landholders and for neighbors.

Expropriation of landholdings for public purposes is legally possible. As mentioned above, some of the projects do not start in a timely manner to use of the expropriated holdings. In such cases, the former land holder informally continues the use of the expropriated land until the realization of the projects.

Some infrastructure projects, such as power line construction, usually pay compensation for land along the whole line. Normally, the previous landholders informally continue cultivating annual crops under the power line. The power line authorities only act, when they observe some interference to their project. So, the power line projects are examples of partial utilization of the expropriated holdings. It is said partial, because the previous landholders are not allowed to plant big trees or to make any type of construction.

Road authorities expropriate land for the road itself and for some additional free land along the road line. The previous landholders usually continue to cultivate the unconstructed area informally. This is also partial use, as the previous landholders are allowed to use only parts of their previous holdings until the land is required for the development of the road authority. In contradiction to the formal setting, there is recognition of latent rights (both as a benefit and an obligation) in the informal setting. Therefore, in the informal setting latent rights exist both, as a benefit and as a burden.

According to the findings of the field study, nearly everyone in the rural areas is the owner of the house he or she is living in. The villages are mostly formed by very closely blood related people with strong social relations. People from outside the village are not capable and not willing to buy houses due to the difficulties to assimilate with the villagers. As a consequence to the limited market, rural houses have less value as collateral for formal as well as for informal credit organizations.

The thesis outlined that cultivated landholdings are used as collateral in the informal credit market. The creditor can either sell the land in the informal market or to use it in the case of default. Therefore, unlike the formal system, collateral right exists as benefit as well as obligation to a holding right in the informal setting. The beneficial right is related to the possibility of using the property as a guarantee to get loan from informal lenders.

In the LCDM public advantage is defined as advantage for ownership. For the purpose of the discussion of the informal setting in this paper, the term public advantage has to be modified to social advantage. The social advantage is the added asset and/or benefit to the holding right due to social relations. The social benefits are supposed to serve the common goods of the local society. The individual holdings are the beneficiaries of the general outcomes of the sanctions as members of the local society. Maintaining peace and order in the local society by establishing accepted norms is the purpose of social sanctions. The individual holdings benefit from implemented sanctions since the objective is the common good of the local society.

The members of the local society are responsible for supporting the elderly. Landholders are benefited by negotiation roles of the elderly. The church also has a significant role in strengthening the social bond. The local society members are responsible to attend church ceremonies, where most of the conflicts are resolved. The church defined some days to be off working days. These holidays are the time for the landholders to gather, to share information, and to solve different kinds of problems. Market places are also used for communication and passing information between each other.

Social groups are formed to facilitate practices and services that hardly can be performed on individual basis. For example, groups are responsible for harvesting at peak seasons, for organizing funeral ceremonies, and for creating local saving institutions. The group formations are indirectly related to holdings by simplifying the burdens of individual landholders.

Property rights as a set of social rules are tools for the effective use of scarce resources. Scarce resources are often subjects for conflicts. Individual landholders benefit by reduced litigation in case of conflicts as a result of the social sanctions.

The common pool resources are in most cases governed by the by-laws of the local societies. Sanction of the society on improper use of common pool resources by local society members is beneficial. So, for example, grazing lands on swampy areas only can be used efficiently for grazing during the dry seasons. As a result of social sanctions, these areas are protected. The protection of valuable trees by societal rules has similar advantage for the landholders. The exploitation of community forests for traditional medicine products, farm implements, construction material, fire wood, and the like is mainly managed by social rules.

Free grazing and herd management is another example, where land and individuals benefit from the existence of social sanctions. The labor required for keeping animals is drastically reduced if groups are formed. Similarly the labor required for the protection of crop lands from wild and domestic animals is reduced, when the whole adjacent fields are covered with crops. Groups for keeping herds on a rotational basis are formed by landholders in the neighborhood.

Border strips are left between neighboring landholders as boundary marks. Strips are protected by traditional rules from cultivation. The fodder produced from such strips is specifically given to the oxen used as draught power. The individual landholders are benefited by balancing the fodder requirements of their animals with crop production. The border strips are also used as a bound for erosion control contributing to the productivity of the individual holdings. Therefore, public advantage, after being modified to social advantage, can be seen as a beneficial right for the holdings in the informal setting.

The informal setting is not regulated by the state or its representatives. So it is difficult to assume public regulation in the informal setting. The local society has its own sanctions on the individual holdings that are important for the common good. The protection of common pool resources, the water ways and cut of drains, the area closures on steep slopes, the protection of selected tree species, the plantation of hedge rows along the border lines, the protection of the natural forests in and around the church compounds are some of the examples of culturally approved practices reported during the survey for the common good of the society.

The sanctions to enforce social rules are multi-layered and dependent on the level of violation. Usually, for the first violations advices and warnings are given by the influential members of the society. Labor contributions and monetary payments for the affected parties are second level of punishments to enforce sanctions. The third level of punishments contains exclusion from herd management membership, taking away farm implements temporarily, exclusion from social activities and the likes. Key informants also reported that in case of continuous violation of the rules sometimes the local society burns the house or slaughters the animals of the breaker of the social rule. It has to be said that most of these punishments are against the formal law.

The social sanctions implemented by the local society are not uniform in all the sample sites. The sanctions are dependent on the socio-cultural context of the particular society and furthermore, these sanctions are dynamic in nature. The same local society can implement different sanctions for the same thing according to different situations. The sanctions are targeting the common good of the local society. The sanctions shall be implemented even if they are obligations on individual holdings. The landholder will protect the tree even if the tree harbors birds that can affect his crop. He has to maintain border strips even if they harbor rodents and if he loses some land. The landholder has to attend social ceremonies and meetings even if they are not directly relevant to him. The landholder has to respect the order from the informal local leaders and he has to help the weak, specially the elderly.

Some of the social activities, such as celebrating the holidays, attending the church ceremonies and prayers, hosting guests, and helping neighbors while organizing big festive events, are not directly related to the holding right. But the landholder is expected to fulfil all the social requirements to be considered as an active member of the local society.

The social sanctions have similar effect to the informal setting as the public regulations in the formal setting.

5.3 Points to be Considered During Preparation of Cadastral and Registration Proclamation for Second Level Certification Program in Ethiopia

The tool preparation for the development of cadastral and registration proclamation was guided by the results of systematic evaluation and description of formal and informal rural land administration settings. Cadastral and registration law is identified as a gap during systematic evaluation of ANRS rural land administration system and legal system development was recommended as a solution. The importance of the inclusion of the practices of the informal setting during law making is another important conclusion made during the description of both, formal and informal rural land administration settings in ANRS.

One of the tools in this thesis is a collection of guiding principles for law development. The guide is prepared to advice the law makers and professionals during the development of cadastral and registration proclamation. The tool aimed at proposing the important provisions to manage different interests on land.

5.3.1 Users' needs and interests on land

Spatial description of land administration systems deals with the *'where and how much question'*. A proper spatial description can minimize the frequency of boundary conflicts. Of course land related conflicts cannot be totally avoided by any system. The conflict rate in ANRS is still high. Boundary lines in the current Ethiopian land administration system are still imaginary lines, where the right of one land holder ends and the other right begins. Boundary conflicts are indirect indicators for the need of cadastral surveys.

Accuracy requirements can be based on the type and the nature of boundary conflicts. The study focuses to use the most prominent accuracy requirements as input for the tool. A accuracy need assessment was conducted through individual interviews, through group discussions and through expert panels. The needs were categorized according to the legally recognized holding types in ANRS: individual holdings, communal holdings and state holdings. The survey confirmed that each holding type has different accuracy requirement.

The accuracy demands of individual, communal and state holdings were reported by users and crosschecked by questioning individual land holders, how much boundary shift they tolerate with their neighbors. According to the findings the average accuracy demands to be included in the registration and cadastral proclamations are:
- for individual holdings: $\pm$ 0.2-0.5 m;
- communal holdings: $\pm$ 2-3m; and
- state holdings: $\pm$ 5-10 meters.

There is a consensus by all respondents on the importance of developing a cadastral procedure that can guide and standardize the implementations of cadastral projects in the country. The experts' panels identified the main points to be included in the cadaster and registration

proclamation. According to the experts panels' and the results of legal system review registration and cadastral proclamation should consider the following categories:
- general provisions;
- provisions to consider private interests on land;
- provisions on management interests of the state;
- provisions on access interests of the state;
- transitional and concluding provisions.

The main categories of cadastral and registration proclamation are identified based on the interests on land. General provisions and transitional and concluding provisions will be supportive provisions to the three main interests on land (individual interests, government access interests, and government management interests).

The major interests on land in Ethiopian are attached to the holding rights (individual, communal and state holding rights). Holding right is the link between the subject, right holder, and the object, land (Shibeshi, Fuchs and Mansberger, 2014). The relationship describes the rights, restrictions and responsibilities (RRR) on land. The aim of provisions on restrictions and responsibilities are protections of public advantages and protections of the rights of neighboring land holders. The rights mainly focus on the protection of the benefits of landholders. The optional provisions for each category are documented in the following subsections according to the above mentioned categories.

5.3.2 General Provisions

General provisions have to answer cross cutting issues and they have to serve general purposes. These provisions specifically target on options dealing with proper implementation and functioning of the proclamation. Some of the options to serve the general purpose in the proclamation are listed below.
- Preamble, objective of the proclamation and or introductory section;
- Short title of the proclamation;
- Definition of important words and phrases;
- Registration type (title or deed);
- System of first surveying and mapping (sporadic or systematic);
- The system for unique parcel identifiers;
- Geodetic control points and connection to the national grid;
- Explicit recognition that the register reflects the ultimate legal status of the registered immovable properties and rights in immovable property;
- Provisions creating a unique cadaster and land registration system;
- Basic provisions for the cadaster specifying what cadaster includes - textual and graphical data about basic property units, buildings and in special cases utilities;
- Provisions about cadaster units and their designation;
- Provisions about connection of measurements to geodetic reference network;
- Defining of cadastral measurements; maintenance and modernization of geodetic reference; networks, survey needed for property formation and other surveying activities related to maintenance of an updated cadaster;
- The range of the relative and absolute accuracy demands for different holding types (as outlined in Chapter 5.3.1);

- Identification of the taxable unit (parcel, holding).

According to individual respondents, the source of conflict is higher correlated with pride and social status than with land value.

5.3.3 Provisions to Consider Private Interests on Land

The holding right in Ethiopia includes individual (private), communal, and state holdings on land. Individual holdings are the landholding rights given to natural persons, to legal persons such as firms, family holdings (common holdings), organizational holdings, etc. The provisions to address private interests on land are categorized into:

1. Provisions to describe rights of the land holders;
2. Provisions describing the responsibilities of the land holders;
3. Provisions to define the restrictions on the land holders.

5.3.3.1 Provisions to Describe the Rights of the Landholders

The right to hold land emanates from the constitution of Ethiopia. The right to use, to transfer and to dispose the property developed by the individual's labor, creativity or capital is given to individual land holders. Land can be transferred together with developed property. But land in Ethiopian condition cannot be subjected to any form of sale or exchange.

Attaining tenure security is the main objective of the protection of landholder's rights. The optional provisions to describe and protect landholder's rights are listed below.

- Mandatory registration of immovable property and their rights;
- Provisions for determination of fees to be charged for cadaster and land registration;
- Provision regarding the right to obtain a certificate of the last recorded data in the registers against payment of fee;
- Standard title certificates;
- Provisions on application for registration with standardized processes, forms and list of documents required;
- Guidelines, examples and standardized forms for registration in the cadaster and land registers;
- Receipt book and pending applications;
- Proof of identity;
- Replacement of lost certificates;
- Provisions regarding time limits for registration and effects of violation of time limits;
- Provisions regarding the effect of registration;
- Provisions specifying reasons for rejection of registration and procedures for rejection;
- Contents of instruments for transfer, exchange, lease, mortgage, and easement;
- Title registration of indefinite or very long-term rural and urban leaseholds;
- Title registration of a right of user other than indefinite or very long-term rural and urban leaseholds and of easements.
- Provision of a specific list of rights and encumbrances that have to be registered;
- Provisions of property formation without regulative plans;
- Definition of property formation i.e. subdivision and amalgamation;

- Provisions for property formation, boundary determination, and boundary adjustment;
- Provisions for the registration of will and gift;
- Land rights and restrictions valid without registration.

5.3.3.2 Provisions to Describe the Responsibilities of the Land Holder

Responsibilities to use land in a sustainable manner emanates from the constitution. The landholders, when exercising their rights, have to protect long term and short term benefits of the society. Clearly defining and enforcing provisions on landholders' responsibilities is one of the main tools for sustainable development. The concept of sustainable development is based on wise land resource utilization. Land resources are supposed to serve many generations. Provisions to describe and enforce responsibilities of land holders are:

- Registration of shares in specified land rights;
- Consent of mortgages obtained before subdivision is registered;
- Time limit for filing of registration application after completed property formation study;
- Time limits for filing of application after completed boundary determination and boundary adjustment studies;
- Land use planning provisions;
- Development control provisions.

5.3.3.3 Provisions to Define the Restrictions on the Land Holder

The aim of restrictions on land interests are based on the principle of protection of the public advantage and of protection of rights of neighboring landholders. In addition to natural resource conservation, environmental protection, and sustainable use of natural resources, the restrictions play a vital role in maintaining social relations and in avoiding conflicts as a result of competition for scarce land resources. The ability of land administration systems to properly enforce restrictions on land can create trust on the system. Trust worthiness (by both the landholders and the public at large) is a major requirement for long term investment and efficient land market. Provisions to define restrictions on land holders should aim to bring trust on a system. The provisions to be considered in the sub-section include:

- Registration of mortgages;
- Registration of declaration concerning a fixture;
- Provisions including a list of overriding interests or rights and restrictions that are valid whether or not they have been registered;
- Penalties;
- Short-term leases, if any, that do not need to be registered;
- Provisions to ensure that property formation is in accordance with zonal plans or other decided plans or regulations;
- Provisions that no other interest in an immovable property can be registered until ownership or indefinite or very long-term rural and urban leaseholds of the immovable property has been registered;
- Provisions defining size, shape and land cover/land use of parcels in accordance with involved land holders (boundary negotiations);
- Provisions about proper land use.

5.3.4 Provisions on Management Interests of the State

Rural land administration is a tool to manage humankind to land relations. The relations said to be formal, if they are enforceable by formal institutions. The relation can also be customary or informal, if management rules are created by custom and enforced by customary sanctions. The management interest of the state refers to the management of formal humankind to land relations. The management interests of the state are necessary for the society to exist as a sovereign and unified socio economic unit. The target of the provisions on management interests of the state is equitable sharing of scarce land resources. Scarcity usually triggers conflict. The weaker parties in a society will be disadvantaged by unmanaged land resource utilization. Protection of the legal rights of the weaker parties and creating fair resource utilization rules is the objective of provisions on management interests of the state. The optional provisions to be considered in this section are:

- The guarantee of quality of data;
- Provisions for correction or deletion of cadastral and/or registration data;
- The extent of legal liability for the accuracy of data;
- The extent of rights of privacy over land and property information;
- The provisions on alterations to entries in the registers;
- Definition of the responsibilities of registration officials and rules governing the delegation of powers;
- Court and traditional arbitration committees jurisdiction over claims, disputes and appeals;
- Defining of inspection and provisions regarding cadastral activities by licensed surveying companies or surveyors;
- Provisions regarding the effect of registration;
- Provisions on licensing of surveying companies and surveyors to do cadaster surveying including requirements for license, revocation of license, administrative appeal of license decisions;
- Listing and defining of data, both for basic property units and buildings, to be recorded in both textual and graphical parts of the cadaster. Be aware of the necessity of data before entering them on the list;
- Provisions stating the legal status of electronic and/or written records; the procedure to upgrade from one to the other and which one has the legal power when;
- Provisions specifying grounds for rejection of registration and procedures for rejection;
- Provisions regarding time limits for registration and effects of violation of time limits;
- Registration of rural Kebeles in the boundaries of urban administration;
- Registration of national parks and state forests;
- Education/Knowledge of Land Administration Staff;
- Provisions about the design of cadastral maps.

5.3.5 Provisions on Access Interests of the State

Access interests of the state are important tools for managing land to humankind relations. The access interests include not only physical access interests, it is also comprehending land information and land data access provisions. The provisions outlined in this section try to balance the privacy need of individuals and the public right to get access to both, to service areas and to land information. To this end, the following provisions are proposed:

- Provisions regarding the registration authority's review of submitted documents and the setting of deadline for completing documentation;
- Provisions regarding acquirements of data for public and private sector use;
- Provisions regarding the registration authority's liability for damages arising from errors, including negligence;
- Provisions specifying when and how the state will be liable for errors in the registers;
- Provisions for administrative and judicial appeal of decisions by the registration authority;
- Government institutions obligated to deliver files requested by registration authorities;
- Provisions establishing an assurance fund and detailed instructions about its operation;
- The ownership of data and the copyright to data within the registers and thus benefit from its sell and use;
- The coordination and cooperation regarding data collection and storing of data;
- Institutions obligated to deliver files requested by registration and cadastral authorities;
- The pricing of data;
- Provisions allowing surveyors access to land during surveying activities;
- Provisions for protection of boundary marks.

5.3.6 Transitional and Concluding Provisions

Proclamation of any kind is part of the country's legal system. The transitional and concluding provisions are used to create the link to an existing legal system and defining the responsibility of major stakeholders for effective implementation of the proclamation.

Designation of a single authority responsible for the integrated cadaster and land register and with the authority over the overall performance of the registration as well as a statement that registration at other institutions does not affect cadaster and title registration;

- Provisions for correction or termination of register data;
- Provisions for archiving of electronic and/or written records;
- Provisions for reconsideration and review of decisions of registration authorities;
- Designation of local cadaster and land registration authorities, and
- Definition of responsibilities of the Land Administration and Use Committees;

5.4 Geodetic Control Points

Geodetic control points in the ANRS are evaluated for their suitability for the second level certification program in the area. The evaluation is limited to the criteria availability, point description, and accuracy criteria in comparison to the needs of the intended cadastral surveying and second level certification program. The study is also limited to the two main geodetic networks available in the region, namely the EMA (European Mapping Agency) and to the AM network (established by BoEPLAUA). Based on the findings and based on the reported high demand of GCPs in ANRS to be used for the planned second level certification program, cost effective method facilitating the creation and the densification of Ground Control Points (GCPs) was part of the current study.

5.4.1 Network Established by BoEPLAU (AM network)

The geodetic network established by BoEPLAU (called AM network) has 42 points distributed all over ANRS. All AM network points are generally in good condition. The points are established in close range to all weather roads. They all are easy to locate and use. During the field observation only one (located in Debretabor) amongst all 42 points was identified as non-usable because of a quarry excavation in the area. Point distribution of AM network is depicted in *Figure 11*.

Figure 11: Point distribution of AM network

5.4.1.1 Accessibility and Status (AM network)

The control points in AM network were created on very stable bedrock that cannot be easily destroyed or moved. The points are marked by painted and drilled triangle and the point itself is realised by a drilled hole in the middle of a triangle. Point names are also painted and drilled, as the names written with continuous point drills stay longer than the paints. *Figure 12* shows the picture of AM points.

Figure 12: Picture of an GCP in the AM network (Source: Miskas & Molnar, 2010)

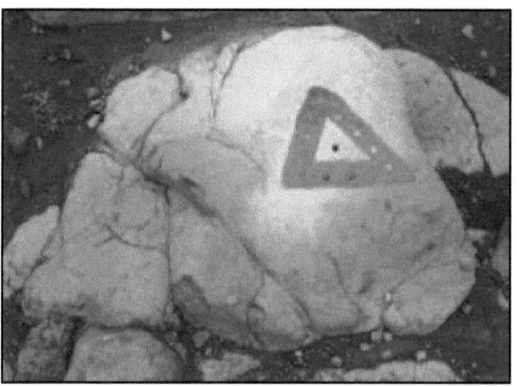

5.4.1.2 Point Description (AM network)

The unique identification number of AM grid points has four features. The first part is '*AM*' to symbolize that it is located in ANRS. The number next to '*AM1*' denotes that it is first order point and the following number ('*AM1-5*') is the unique code for the specific point. This number is followed by the name of the area ('*AM 1-5 Gishe abay*') (Miskas & Molnar, 2010). *Figure 13* shows details of the template and *Figure 14* is an example for the location of AM points.

A pre-defined template for point descriptions was compiled for the AM network to enable an easy use. The point descriptions include pictures of the surrounding area, the location of grid point in a topographic map (1:50,000), hand drowns sketches, textual descriptions and point coordinates.

The description is easy to use because it contains current fractures as a reference and it is supported with pictures. The location on the topographic map serves as guidance for reaching the area. In general, point descriptions of AM network are very convenient for the planned second level certification program.

Figure 13: Template for an AM grid point (Source: Miskas & Molnar, 2010)

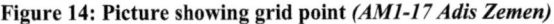

Figure 14: Picture showing grid point *(AM1-17 Adis Zemen)*

5.4.1.3 Accuracy (AM network)

The accuracy level of AM points was determined by comparing the outlined coordinates with coordinate measurements of a precise point positioning (PPP). The variation between East and North values of AM points and PPP were calculated. The correlation between East and North values for the same points is insignificant (NRCANGSD, 2012). The variation is said to be insignificant, because the root mean square error (RMSE) in East is ±0.088 m and in North ±0.084 m. RMSE is a measure to describe the variations between point coordinates of PPP and known coordinates of AM point. RMSE is estimated by squaring the calculated differences between PPP and AM points, adding those together, dividing that by the number of total number of samples, and finally taking the square root of that result.

The RMSE of both north and east coordinates lies below the required accuracy for the planned cadastral project. The table is attached in the appendix for details. The correlation is significant between PPP East and AM East and PPP North and AM North at 0.01 levels. The correlation coefficient of AM East and PPP East as well as for North values of AM and PPP is 1.000, which means that the correlation is significant (see *Table 10*).

Table 10: Correlations of AM and PPP measurements

		PPP E	PPP N	AM E	AM N
PPP E	Pearson Correlation	1	.019	1.000**	.019
	Sig. (2-tailed)		.855	.000	.855
	N	96	96	96	96
PPP N	Pearson Correlation	.019	1	.019	1.000**
	Sig. (2-tailed)	.855		.855	.000
	N	96	96	96	96
AM E	Pearson Correlation	1.000**	.019	1	.019
	Sig. (2-tailed)	.000	.855		.855
	N	96	96	96	96
AM N	Pearson Correlation	.019	1.000**	.019	1
	Sig. (2-tailed)	.855	.000	.855	
	N	96	96	96	96

**. Correlation is significant at the 0.01 level (2-tailed).

The AM network is connected to Addis and Jima core stations (www.ipg.tu-darmstadt.de/en/adis_igs.html). The data from the two stations were converted to a common epoch using a velocity factor. It is also transformed to common reference frame using seven parameters. The velocity factor used for conversion was generated using NUVEL 1A model for plate motion calculator developed by UNAVCO. The velocity factor of each AM grid point is attached in the annex.

5.4.2 Network Established by Ethiopian Mapping Agency (EMA Network)

Ethiopian mapping Agency (EMA) is the responsible authority for the establishment and mapping of geodetic points in Ethiopia. The Ethiopian Mapping Agency is using Clarke 1880 spheroid and UTM projection in metric unit based on Adindan Datum. The Blue Nile datum of 1958 created with the support of U.S. Department of commerce coast and Geodetic survey for Ethiopia was the first datum of Ethiopia. The origin of the geodetic work was in southern Egypt, south of Lake Nasser, at station Adindan.

Until now about 80% of the country is covered with primary and secondary geodetic control points with an approximate interval of 50km. There is also a plan by EMA to density the control points. The distribution of the planned and existing ground control points by EMA is shown in *Figure 15*.

The plan of EMA shows that the future focus will be on the densification CORS and first order points all over the country. But even after the construction of all the planned points, the density of EMA control points will be very far from enough for undertaking cadastral surveying that requires on average one point in every kilometer square.

Figure 15: Existing and planned distribution of ground control points by EMA (source; Sultan, 2012)

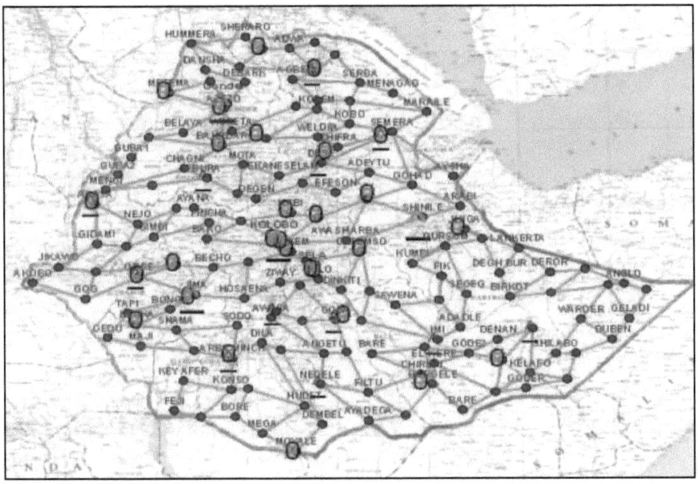

5.4.2.1 **Accessibility and Status (EMA network)**

The monuments of geodetic points created by EMA are very old and not properly maintained. Point descriptions are very poor and obsolete to guide a user to find them. The coordinates of the points are very difficult to rely on, as some of them are moved, destroyed or significantly damaged (Miskas & Molnar, 2009).

In addition to scrubby distribution of the geodetic points, they are located mostly on inaccessible hill tops. Nearly all of EMA points were created on very remote mountain tops and on cliffs that are not accessible. The control points are not reliable, as they are rarely attended and maintained since many years (Palm, 2006).

According to the findings of the current study, 10 of the randomly selected 30 points were destroyed. Only 9 points were easily accessible and located in less than 5 km distance from all-weather road (see *Table 11*).

Based on the field result it can be stated that EMA points are located on very difficult positions and many of them (33.3%) are destroyed. Therefore, based on this criterion they are of limited suitability for the planned second level certification program of ANRS.

Table 11: Availability and status of EMA points

#	NAME	E	N	H	Availability	Remark
1	AFKERA MICAEL	392104.90	1119864.50	2414.608	No	
2	ANBISSI	298558.695	1210549.41	2922.062	Yes	3km from main road, quite steep hill
3	BACHEMA	302575.447	1265978.38	2146.298	Yes	Hill/Mount 2km from main road, correct name
4	BEMRE MARKOS ASTRO	362004.342	1141346.85	2450.9	No	
5	BRADY	288072.548	1194088.33	2631.431	No	
6	CHOKE	332476.790	1227924.59	3298.5	Yes	Hard to reach, needs climbing
7	CURVE	362516.327	1089253.37	951.5	No	
8	DAR	323959.73	1282424.94	1787.771	Yes	Ghion
9	DEBET MARIAM	398904.018	1146026.78	2745.162	Yes	Close to road on a hill
10	DELMA MICHAEL	338263.981	1159151.98	2380.016	Yes	4-5 km from road
11	GHIETEM	375597.925	1131336.85	2547.7	Yes	Accessible, road exist according to topo map
12	GONDAR ASTRO	328140.344	1384614.19	2137.8	No	
13	GONDAR ASTRO azimut mark	329794.633	1384208.74	1964.153	No	
14	GUDAR	352735.558	1090550.39	1313.3	No	
15	GUNGHI S W BASE	388764.370	1128352.83	2455.3	Yes	Accessible, crop field, right outside village
16	KES MEDIR	340623.555	1183446.66	2595.865	No	
17	KURBEYAL	377764.715	1189537.58	3899.096	Yes	No road, very steep
18	MANGESTU	377820.703	1159163.19	3066.8	Yes	High elevation
19	MENGISTU KIDANEMEH ERITE	399905.075	1169045.45	2710.075	Yes	10 km from road on farm land
20	MOTA	371715.063	1225082.87	7853.67	Yes	Wrong height, on topo 2394 m
21	NILE	366140.475	1242671.46	2206	Yes	Point name is Atsed, hard to reach
22	OMATA	358876.22	1137558.41	2465.3	No	

#	NAME	E	N	H	Availability	Remark
22		6				
23	ROBIT (shafo mariam)	375620.402	1131185.82	2545.65	Yes	Hill, 100m, 2km from main road, small road to point
24	SAGADO	336034.648	1214500.90	3298.2	Yes	Hard to reach, needs climbing
25	TANA	338855.038	1310268.51	1991.9	Yes	Peak
26	TIF	380223.992	1205955.65	2658.007	Yes	Very close to road, about 100 m
27	UABI	387636.853	1128844.34	2531.4	No	
28	UATZAU	329296.130	1120186.42	2126.8	Yes	Far from road, steep canyon
29	WORKE KIDANEMEHIRET (RIDGE)	358705.201	1191076.75	3412.182	Yes	Close to road about 60 m height difference between road and point
30	YEMISTINA	377260.955	1118284.59	2373.803	Yes	On road (dirt road)

5.4.2.2 Point Description (EMA network)

Most of the EMA control points were established before 50 years. As the land use and other features used as indicators for the description of EMA points changed drastically within this period, it was difficult during the field work to find a reference point guided by its description. The documentation of the point was influenced by the means of transportation during the survey time. At the time of implementation it was necessary – due to measuring methods – to establish the reference points on high hills (see *Figure 16*). These were normally inaccessible by car, as the road network was poor. The use of helicopters was common and that is evidenced in the descriptions. For example, most distances were estimated based on the flight time with helicopters to a given direction.

Figure 16: Picture showing Adet EMA point

Figure 17: Example for older EMA point descriptions

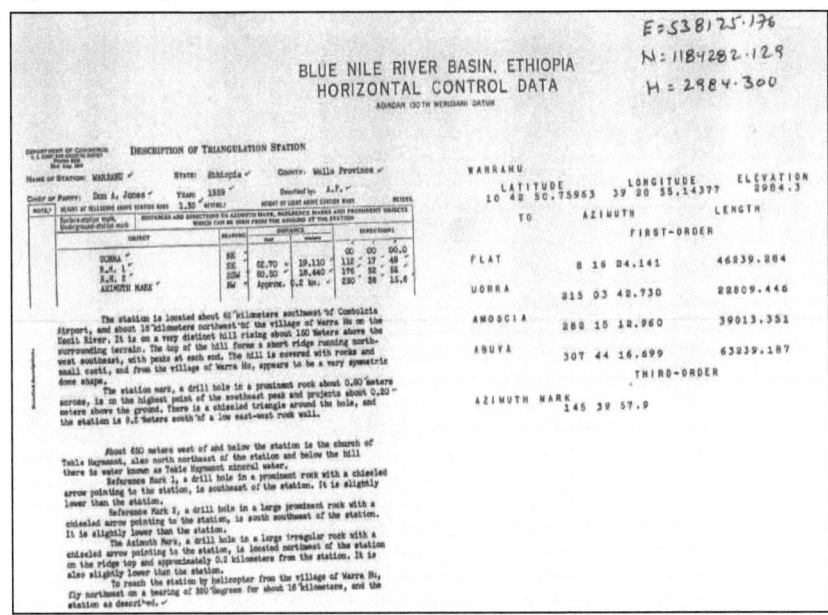

Illustrations, such as pictures or location maps were not available. The only remaining important information in the description sheets of EMA geodetic points are the coordinates of the point. Due to the fact that there is no agreed template for point description and because of the inclination of surveyors to base new documentations on the previous one, the descriptions created recently are also poor (see description examples on *Figure 17* and *Figure 18*)

Figure 18: Example for newer EMA point descriptions

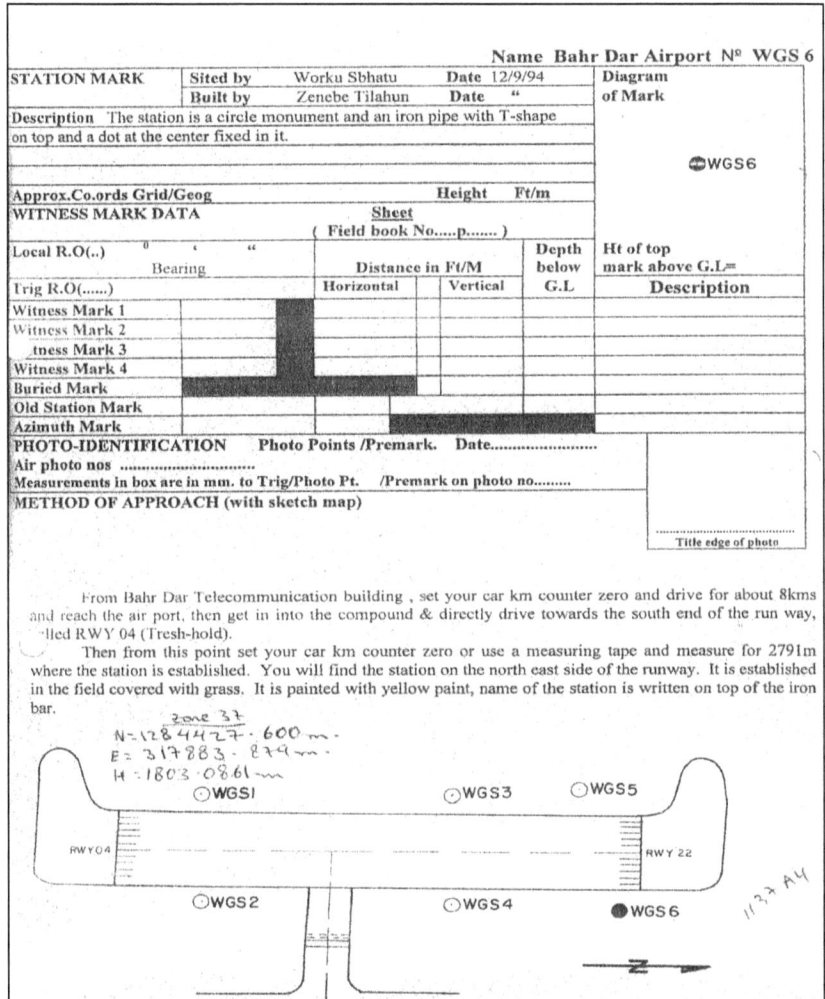

5.4.2.3 Accuracy (EMA Network)

The accuracy level of EMA points is compared with the results from precise point positioning (PPP) that is on line positioning service by Geodetic Survey Division. The variation between East and North values of EMA point and PPP calculated using data from 15 sessions on ten points. The root mean square error is the measure of accuracy level. According to the data analysis result, the average shift in North is -0.20m and in East the shift is +1.11m (*Table 12*).

Table 12: Comparison of EMA points and their PPP calculated equivalent

Point code	PPP coordinates		EMA coordinates		△ East	△ North
	East	North	East	North		
BHP 3	320275.05	1281864.55	320275.31	1281863.74	-0.26	0.82
BHDRa	324149.98	1278374.46	324149.71	1278373.76	-0.27	0.70
BHDRb	324149.92	1278374.46	324149.71	1278373.76	-0.21	0.70
BDR 1	319567.01	1280592.38	319567.23	1280591.70	-0.22	0.68
BDR 2	326980.01	1282957.93	326979.73	1282956.77	-0.29	1.16
BDR 3	324441.65	1278137.12	324441.37	1278136.64	-0.27	0.48
BDR 4a	322901.64	1284109.55	322901.74	1284108.46	-0.09	1.09
BDR 4b	322901.66	1284109.55	322901.74	1284108.46	-0.08	1.09
ADET ROAD	1278167.05	324054.85	1278166.93	324056.88	-0.13	2.03
debanke	1281657.21	320180.78	1281656.94	320182.50	-0.26	1.72
Gion	1282424.71	323961.16	1282424.94	323959.94	-0.23	1.22
Gion 1	1282424.71	323961.21	1282424.94	323959.94	-0.22	1.28
Gion 2	1282424.74	323961.25	1282424.94	323959.94	-0.20	1.31
ADET ROAD 2	1278167.12	324055.64	1278166.93	324056.88	-0.19	1.24
Adet	1249892.54	333722.93	1249892.47	333721.76	-0.07	1.17
				Mean △	-0.20	1.11

The correlation coefficient to show the relation between EMA and PPP calculated coordinates of the same points is calculated and the result show the relation is straight and constant.

Table 13 shows the correlation result of the comparison is both East and North values of EMA and PPP measurements is attached. Based on this result we can conclude that the accuracy level of EMA network is not as good as AM network to be used for the planned cadastral surveying project in ANRS of Ethiopia.

Table 13: Correlations of EMA and PPP measurements

		EMAY	EMAX	PPPY	PPPX
EMAY	Pearson Correlation	1	-.939**	1.000**	-.939**
	Sig. (2-tailed)		.002	.000	.002
	N	7	7	7	7
EMAX	Pearson Correlation	-.939**	1	-.939**	1.000**
	Sig. (2-tailed)	.002		.002	.000
	N	7	7	7	7
PPPY	Pearson Correlation	1.000**	-.939**	1	-.939**
	Sig. (2-tailed)	.000	.002		.002
	N	7	7	7	7
PPPX	Pearson Correlation	-.939**	1.000**	-.939**	1
	Sig. (2-tailed)	.002	.000	.002	
	N	7	7	7	7

** Correlation is significant at the 0.01 level (2-tailed).

5.4.2.4 Long Duration GNSS-measurements

Groups of five GNSS (Global Navigation Satellite Systems) measurements on eight different points for the duration of one hour, two hour, three hour, four hour and five hour, were carried out to check the effect of time on accuracy improvement. Based on the reference, the average mean square error for East and North is calculated and presented in *Figure 19* to show the magnitude of the error for different measurement durations.

Figure 19: Root mean square error for different observation times [in ± m]

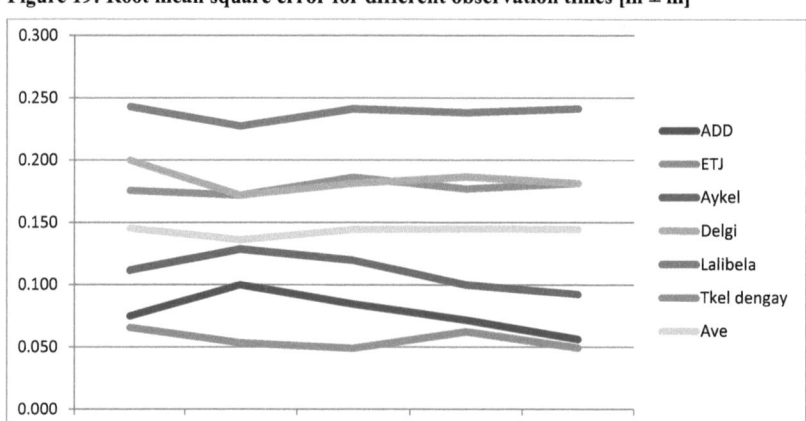

No apparent data variation is observed after one hour measurement. The results obtained after 30 minutes of static measurement by dual frequency GNNS equipment were very stable. The correlation coefficient is also 1.000 showing that the relation is straight and constant. *Table 14* documents the achieved correlation result of the comparison. East and North values of different duration measurements with a control are attached.

Table 14: Correlations between different time length static measurement results

		Reference	one hr	two hr	three hr	four hr	five hr
Reference	Pearson Correlation	1	1.000**	1.000**	1.000**	1.000**	1.000**
	Sig. (2-tailed)		.000	.000	.000	.000	.000
	N	11	11	11	11	11	11
one hr	Pearson Correlation	1.000**	1	1.000**	1.000**	1.000**	1.000**
	Sig. (2-tailed)	.000		.000	.000	.000	.000
	N	11	11	11	11	11	11
two hr	Pearson Correlation	1.000**	1.000**	1	1.000**	1.000**	1.000**
	Sig. (2-tailed)	.000	.000		.000	.000	.000
	N	11	11	11	11	11	11
three hr	Pearson Correlation	1.000**	1.000**	1.000**	1	1.000**	1.000**

	Sig. (2-tailed)	.000	.000	.000		.000	.000
	N	11	11	11	11	11	11
four hr	Pearson Correlation	1.000**	1.000**	1.000**	1.000**	1	1.000**
	Sig. (2-tailed)	.000	.000	.000	.000		.000
	N	11	11	11	11	11	11
five hr	Pearson Correlation	1.000**	1.000**	1.000**	1.000**	1.000**	1
	Sig. (2-tailed)	.000	.000	.000	.000	.000	
	N	11	11	11	11	11	11

**. Correlation is significant at the 0.01 level (2-tailed).

5.4.3 Recommended Method for the Densification of the Ground Control Point Network

The precise orbits and clocks used by Canadian Spatial Reference System - Precise Point Positioning (CSRS-PPP) remove a large part of the GNSS errors. In addition, CSRS- PPP processing must also properly account for several other effects on the position of the GNSS receiver (Moreno et al., 2011; Seredovich, Irughe & Ehigiator, 2012; AUSPOS, 2012). The comparison of the existing coordinates of control points with the result of CSRS-PPP outlined cm level standard errors both in north and East. Other similar studies also confirmed mm to cm level results at 99% confidence level in different places (Ebner & Featherstone, 2008 ; El-Mowafy, 2011; Seredovich, Irughe & Ehigiator, 2012; Rizos et al., 2012).

The following seven online GNSS post-processing are comparable:
- CSRS-PPP: Canadian Spatial Reference System, Natural Resources Canada
- AUSPOS: Geoscience Australia
- GAPS: University of New Brunswick
- APPS: Jet Propulsion Laboratory
- SCOUT: Scripps Orbit and Permanent Array Center (SOPAC), University of California, San Diego
- magicGNSS: GMV
- Center Point RTX: Trimble Navigation (Gakstatter, 2013).

After considering the existing two networks and comparing them with PPP, the following method is recommended to be used for the establishment of ground control points that can be used for cadastral surveying in the Amhara region of Ethiopia:

According to the current study, AM network is more accurate and current than EMA network. The point description template of AM network is found to be the most suitable for the purpose. AM network is also good in terms of accessibility and availability of points compared to EMA network. The total density of ground control points, of both networks, is very low to support the planned cadastral survey projects and needs to be densified using PPP method. Benefits of using PPP method are: the speed of surveying can be increases as there is no network adjustment; the cost can be decreased as only one base has to be used for creating a point; and no post processing software is needed.

The responsible institutions shall consider upgrading and maintenance of the existing EMA points. The destruction of ground control points monuments have to be avoided. The template for point descriptions of all points in the area during maintenance and updating should be based on AM network point description template.

The main reason for the destruction of the ground control points was the lack of a responsible body for protection. Before the beginning of the implementation of the planned cadastral project all ground control points in the region have to be identified, maintained, updated and recorded as a property of either EMA or BoEPLAU. The list of control points including the description have to be given to the respective district and parish level administrations, so that they can provide proper protection for the control points.

In selecting the locations for new reference points the main considerations should be long term stability, security of the marks, safety of users, and accessibility (Rui & Salah, 2013). The technology for point establishment some 50 years ago requires inter visibility between two or more points for measurement (Mugnier, 2003). Nowadays, with the advent of GNSS equipment the importance of visibility between control points is minimized.

5.5 Selection of Cadastral Survey Methods

The commencement of cadastral surveying in progressive land administration systems normally requires considerations, investigations and decisions at several stages. It has to be checked, if – additional to the existing first level of certification – cadastral surveys and maps are required to fulfill the objectives of a land administration system and in which way cadastral maps can produce added value to other land-related issues. Attention also has to be given to financial issues. The right time to start cadastral surveys is also an important consideration in the pre-implementation phase. And finally, the proper cadastral survey methods have to be selected to assess in cost-effective way spatial information for different kinds of holding types.

5.5.1 Necessary Considerations before the Implementation of Cadastral Survey Projects

The role of cadastral maps in land administration systems is to answer properly the questions of the location and size (area) of the land holdings. The importance of precise information about location and size of the property is manifold in the continuum of the development of land administration systems. The need for cadastral maps is dependent on the overall economic development and the market system. Mortgage and land sale are examples for the important economic activities that require spatial information of the land parcel.

Considerations required before launching cadastral survey projects can be classified into
- Objective and goal setting criteria;
- Timing and need assessment criteria; and
- Way of working design criteria.

5.5.1.1 Objective and Goal Setting Decision Criteria

This kind of decision criteria deal with:
- The inclusion of cadastral concepts in policy and constitutional provisions;

- The importance given to cadastral surveys in the country or state level development plan;
- The suitability of economic policy and market forces to benefit from land transactions;
- The capacity to administer spatial data at establishment and maintenance stage

Policies and constitutional provisions have to support and acknowledge the importance of cadastral survey projects. They are important, but will become toothless unless they are supported by lower level laws such as proclamations, regulations and directives. In ANRS the policy and framework level laws are fulfilled, but the supportive detail of cadastral and registration proclamation is missing. The ANRS therefore has to prepare and enact a cadastral and registration proclamation before launching the second level certification program. Since cadaster is the surveys of legal boundaries, the existence of law for adjudication of rights and proper identification of legal boundaries is important.

The importance given to cadastral surveys in the country or state level development plan is another consideration to decide about the timing of cadastral projects. Priority activities are usually indicated in state level growth and transformation plans. The inclusion of second level certification program in the Ethiopian growth and transformation plan can be seen as a first positive step forward. A very ambitious goal is to issue second level books of holdings of all small scale farmers within the next five years in four regional states (Shibeshi, Fuchs & Mansberger, 2014). The pace of implementation seems to be in contradiction to this schedule. Therefore the ANRS has to update the plan and to take necessary preparatory measures before implementing the second level certification program.

The objective of attaining tenure security for sustainable development was properly addressed by issuing primary books of holdings and by protecting the property rights of landholders in the regional state. The expected added value from second level certification program is the feasibility of transactions and mortgage. Policy measures, such as liberalizing property right transactions and the possibility of lien for small scale landholders, are some of the necessary policy measure to be taken to achieve full benefit of the second level certification program in ANRS.

Technology is an important and demanding input for the establishment and the maintenance of a well-functioning cadastral system. The assessment of spatial data infrastructure covers large part of costs required for running a progressive land administration system (Ali, Tuladhar & Zevenbergen, 2010; Shibeshi, Fuchs & Mansberger, 2013). The strategy to choose appropriate technology to meet the objectives and aims of the progressive land administration system has to focus rather on the needs than on the technical standards. The technology strategy of the ANRS land administration system therefore is leaded by the governing concept to serve the aim and objectives of the system and by minimizing the cost for the establishment and maintenance of the spatial data infrastructure. At the initial stage of the land administration project the use of a spatial component was reduced to bare minimum. Now, the improvement of the geometric information is planned by using cost effective surveying technologies, which had been tested in several pilot projects.

Strong institutional setting is one of the determining factors for realizing the objectives of the progressive land administration system. All in one kind of institutions has proved to be effective for land administration purposes. In the ANRS the land administration is a unique institution, but needs to be reformed. Core businesses processes have to be in line with core land administration functions. Another important factor is a proper mission, vision and strategy being in line with the aim and objective of the land administration system. The ANRS has a well synchronized mission, vision and strategy, but the level of the staff awareness on these documents needs to be increased (Shibeshi, Fuchs & Mansberger, 2013).

5.5.1.2 Timing and need assessment criteria

The readiness of the system is the result of institutional, users, stakeholders, and policy and law readiness to install and maintain well-functioning cadastral system. The right time to launch cadastral projects is, when the main actors of the land administration system are ready and the environment for operating the system is favorable.

One of the main reasons for failure in land administration projects in Sub-Sahara Africa is the introduction of accurate surveying and mapping at the wrong time. Before the commencement of the spatial component the land administration system has to be ready to shoulder the demanding task.

Capacity in terms of human power, the way of working and the materials to establish and manage the spatial data infrastructure is an important issue. Capacity building commences from the identification of available resources and improving them to the level required to attain the envisaged objective. Capacity building programs are necessary to establish and maintain well-functioning cadastral systems. The current institutional capacity is important, but it is also necessary to guarantee the capacity also in the future. Higher learning institutions are the continuous sources for trained human power. Curricula and programs of higher learning institutions must be geared to fulfil the needs of the progressive land administration systems.

The institutional setting of emerging land administration systems with and without spatial component is different. The institutions have to capacitate their staff and open new positions to plan and execute cadastral systems. The technical capacity of the staff needs to be sufficient to implement the selected surveying and mapping method. Necessary materials have to be available for the implementation. In ANRS BoEPLAU elaborated with the support of SIDA a surveying manual, which is used as a guiding tool and training material for surveyors at BoEPLAU.

Budget and financial supply for the planned second level certification program is another important issue. The funding for the implementation can be from government sources and/or from external donors. The financial means have to be allocated in time and enough in amount to undertake the task. Currently there is no clear information, whether the budgetary input needed for the second level certification program in ANRS is ready or not.

It is the public responsibility to put up capital for the first instalment of the spatial data infrastructure. Users of land administration system shall cover the updating and the maintenance costs.

In ANRS the funding of the issuance of both first and second level certificates lies in the responsibility of the regional state. Updating and other land administration services will be conducted on cost recovery principle. According to the findings of the respondents in the individual interview the land holders are willing to cover these costs.

The strategy for the implementation of land administration system in the ANRS specifies the second level certification program as a continuation of the first level certification program. The first level book of holding will be upgraded into second level by adding spatial component on the textual data created during first level certification program. Before the data sets of the first level certification program have to be updated and computerized.

ISLA (Information System for Land Administration) was created to computerize and update the textual first level certification data. The most recent report of BoEPLAU indicated that the records of 1,372,065 land holdings and 6,038,914 parcels in 74 different districts are encoded in ISLA.

The readiness of the national grid and with it the availability of enough geodetic control points is an important requirement to create a cadastral map that is connected to the national grid. Among other benefits, a cadastral map connected to the national grid is the basis for multipurpose cadastral uses (for the details see *Section 5.4*).

The multipurpose use of cadastral systems is influenced by users' readiness. In awareness creation programs the users have to be informed about the potentials of a multi-purpose cadaster. A level of basic knowledge, such as map reading and interpretation, is important for a proper use of cadastral maps. But users also have to be ready to cover the updating and maintenance costs.

The level of education and skill of rural landholders in the ANRS in map reading is at a rudimentary stage. However, landholders demonstrated a skills in orthophoto interpretation during the fieldwork outlined within this study.

According to the respondents of the questioner survey 80% of the users are willing to cover the costs for second level certification program. Similar results (90%) were reported by other studies (Deininger et al., 2008). Decline in willingness to pay was identified only in Oromia and in SNNP regions (Bezu & Holden, 2013).

The objective of second level certification program in the ANRS is establishing a well-functioning multipurpose cadaster. Different stakeholders are supposed to share and use the multipurpose cadastral data. Stakeholders' readiness is a requirement for efficient use of spatial data infrastructure. Readiness is the reflection of the need for spatial data to effectively undertake their businesses and the ability of the infrastructure to integrate them. Stakeholder institutions demanding cadastral data for their daily work are courts at different level, taxation and revenue authorities, banks and financial institutions, development offices, and so on. Currently the stakeholder institutions ANRS are not prepared to provide location based on the outputs of the planned second level certification program.

Also the legal system, including relevant policies, laws and regulations, needs to be ready before full scale implementation of spatial data infrastructure. A cadaster and registration law with sufficient details on the assessment and maintenance of land information is necessary for effective implementation. The ownership and the mandate related to a spatial data have to be clarified. The role and responsibility of different institutions in creating and using spatial data have to be defined. The land administration system of the ANRS was created based on law of the land and the necessary legal system was in place. However, a detailed cadastral and registration law covering ownership, roles, responsibilities, and major stakeholders is still missing.

5.5.1.3 Way of working design criteria

The development of a way of working and of working procedures is a pre-request to launch cadastral projects. The focus of the present day cadaster is not only about fulfilling strict surveying standards. Users' participation and addressing users' needs are also important success factors for proper and sustainable land administration systems.

The following issues have to be considered:
- Users' participation (technical surveys included);
- Support for the weaker parties;
- Conflict management procedures;
- List of missing boundary establishment guideline based on their priority order
- The designation of unique parcel identifiers for both spatial and textual data;

- Accuracy need and data quality for each land holding/ownership type;
- Possibility for upgrading and data migration;
- Possibility for rechecking, appeal and modifications.

The way of working decision criteria deals with the availability and the quality of a procedural guideline for the assessment and the maintenance of spatial data infrastructure.

In ANRS the above listed working and design criteria for the implementation of the progressive land administration system are fulfilled. Users' participation was insured at all stages of decision making (see *Figure 20* as an example). In addition to direct participation users were represented by parish level land administration committees. The legal system has clear provisions for the support of weaker parties. Women rights are protected both legally and in practice. Conflicts are managed locally. The traditional Shemagelewoch shengo (elderly arbitration committee) was established in all parishes and linked to the formal land administration committees. The traditional arbitration committees are supposed to mediate conflicts using cultural means. Once agreed in the traditional setting, conflicts cannot be resubmitted for litigation in the formal system. If agreement cannot be reached using customary way applications the case can be submitted to the formal legal system, and decisions will be given based on the law of the land.

Figure 20: User participation during land certification process

5.5.2 Selection of Proper Surveying and Mapping Techniques

Cadastral surveys shall deploy different methods for the assessment and the maintenance of spatial information. For the first data assessment systematic surveys has to be carried out, as they are more economical. Maintenance surveys for updating the established cadastral system tend to be sporadic.

During the selection process of available land surveying methods the following factors are identified as very important and have to be considered:

- The capability to fulfil the accuracy needs
- The need for skilled labor
- Implementation speed
- Cost effectiveness
- Possibility to re-establish destroyed boundary marks
- Possibility to connect to the national grid
- Room to be combined with other methods
- Potential for diversified use (multipurpose)
- The existing equipment and resources

The introduction of cadastral systems with very high accuracy level requirements is difficult to implement in the developing world. The accuracy needs of a cadastral system have to be based on users' requirements (Williamson, 1981). Different holding/ownership types are related to specific accuracy demands for different holding types. Accuracy requirements are dependent on the type and nature of boundary conflicts. The accuracy need assessment for second level certification program in ANRS was conducted through individual interviews, group discussions and expert panels. The survey confirmed that each holding type has different accuracy requirement. The average accuracy needs reported to be included in the registration and cadastral proclamation is:

- for individual holdings: $\pm$ 0.2-0.5 m;
- for communal holdings: $\pm$ 2-3 m; and
- for state holdings $\pm$ 5-10 meters (Shibeshi, Fuchs & Mansberger, 2014).

Investigations for selecting adequate technologies and proper methods for the assessment of spatial data infrastructure are necessary for the second level certification program. The first stage of certification is already completed without any detailed geometric information. Within the current study boundary points were accessed with the following methods: remote sensing, handheld GPS, total station, and RTK GNSS.

The selection criteria for the establishment of spatial data infrastructure are used as a guiding principle during piloting. The summary of the results of pilot projects and field work are presented in this section.

The lack of clearly visible boundary marks in the rural lands of ANRS is a challenge for remote sensing tools. Remote sensing methods need to be supported with additional field work to fill this gap and to produce maps with features reflecting the ground situation. Pictures of examples of boundary types document the difficulties of boundary identification for remote sensing techniques (*Figure 21* and *Figure 22)*.

Figure 21: Examples of boundary types (furrow)

Figure 22: examples of boundary marks by stone lines and grass strip

Investigating the points assessed with remote sensing techniques and using the RTK GNSS as a reference the following results were achieved: For Angot Yedegera HRSI pilot site with a total of 629 randomly distributed point level measurements the RMSE (Root Mean Square Error) was determined with ±1.4m for Easting and ±1.5m for Northing. Similar comparison made by Gedamu using 213 point measurements in Zenbela perish in the same region resulted in RMSE with ±1.0m in Easting and ±1.1m in Northing (Gedamu, 2009). The studies conducted in the area confirmed that no better than ±1m RMSE using HRSI is possible with current practice in the region. The comparative evaluations made by ELAP in Yeregen perish also concluded on similar findings (ISS, 2013).

Figure 23 shows the sample surveyed areas using HRSI as well as sample points measured by RTK GNSS. Different colors represent the areas surveyed by different team.

Figure 23: Cadastral index maps produced using HRSI and distribution of sample points

The RMSE calculated in Sereba perish located in the Amhara national regional state using 1098 randomly distributed point level measurements using orthophotos produced from aerial photo is ±0.5m in easting and ±0.6m in Northing. *Figure 24* shows the distribution of the sample point measurements on the orthophoto.

Figure 24: Cadastral index maps produced using orthophoto and distribution of sample points

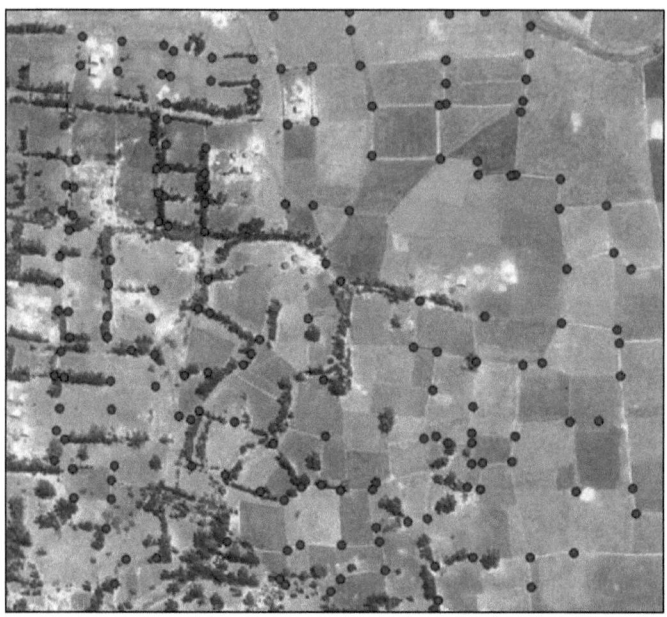

The RMSE calculated using 287 randomly distributed handheld GPS code measurements for Angot pilot site is ±3.7m in Northing and in Easting ±4.4m. Using the same approach, the RMSE calculated using 56 randomly distributed point measurements in the Oromia regional state Illu pilot site is ±2.7m in Northing and ±3.8m in Easting. The RMSE calculated by Gedamu after considering 93 point measurements in Zenbela parish was ±2.2m in Easting and ±1.9m in Northing (Gedamu, 2009).

During the current study also the human power demands for parcel surveying were evaluated. The calculations including the necessary fieldworks and related office work. Using remote sensing tools an average of 40 mapped parcels per day was determined. The average parcels based on investigations of Responsible and Innovative Land Administration (REILA) were 35 parcels per day in Ketta (relatively flat area) and 19 parcels per day in Woleshu (undulating slope) sample sites (Woldeyes & Harris, 2014).

In comparison to remote sensing tools, the average number of parcels assessed by conventional terrestrial surveying was 30.

Cost effectiveness is another criteria to for selecting the proper method. All costs necessary to produce cadastral map are added up and harmonized by the number of parcels surveyed. The cost includes overhead costs, material costs, and running costs.

The estimated average cost per parcel for remote sensing tools is $ 8 for orthophotos and $ 10 for HRSI. The cost per parcel reported by REILA pilot projects was $ 8.5 for orthophotos and $ 11 for HRSI (Woldeyes & Harris, 2014). The cost per parcel for terrestrial surveying tools were split up to the cost for RTK GNSS or total station accurate surveys and to cost for using cheap hand held navigation GPS. The average cost per parcel for RTK GNSS surveys is calculated to be $ 12.6 and the equivalent cost for hand held GPS surveys is $ 6.9. The cost estimated for total station surveys by other studies is 175 Ethiopian Birr that is equivalent to $ 9.7 (Deininger et al., 2008).

However, the cost for handheld GPS survey is presented only for comparison. The method cannot fulfil the accuracy needs and other forthcoming technical selection criteria and therefore handheld GPS techniques are not suited for second level certification program surveying.

The cost and WT (Working time) calculated by ELAP (Ethiopian Land Administration Program) is attached in *Table 15* for comparison.

Table 15: Cost and working hour calculated by ELAP pilot project (Source: ELAP report)

Tools used	Cost		Speed/Survey rate	
	USD/ha	USD/Parcel	WT / ha	WT / Parcel
1-Hand-held GPS	4.5	2.4	34 Minutes	19 Minutes
2-Total Station	6.5	3.5	44 Minutes	23 Minutes
3-HR IKONOS Imagery	12.8	6.9	31 Minutes	17 Minutes

In the case of developed economies return to investment and affordability are very important factors. The capability of good cadastral systems to return manifold the investigation costs is an accepted fact (De Soto, 2000). De Soto even determines a successful cadastral system as a necessary precondition for a successful economy. Affordability is important for developing economies like in ANRS. The method demanding the least costs and fulfilling the minimum requirements of a cadastral system is the first choice.

By selecting the proper method, the possibilities for upgrading and maintaining the land administration system also has to be considered. If the cadastral system has no inbuilt system for updating and maintenance, it will be obsolete very soon. Cadastral maps created by use of remote sensing tools (orthophotos and HRSI) can be easily maintained and boundary marks can be re-established and upgraded with reasonable accuracy level. Maps produced using RTK GNSS and electronic total stations can also be easily updated and maintained. Destroyed boundary marks can be re-established on the same point with sufficient accuracy. Due to the low accuracy handheld GPS cannot be used for upgrading and maintenance of parcel boundaries.

There is no magic bullet that can solve all the mapping challenges and that satisfy all needs of progressive land administration system. Therefore, strategies to combine methods for the assessment of the geometric aspects of a progressive land administration system are needed. The land law of the Amhara national regional state is positive in this regard. It proclaimed that various surveying methods can be applied to meet the different requirements for cadastral mapping.

The case study discussion on maintenance and upgrading selection criteria was carried out only on a conceptual level, as there is no maintenance and upgrading exercise going on in the Amhara national regional state so far.

The availability of trained human power that can implement the selected method is an important factor for choosing an appropriate tool for upgrading and maintenance. Unlike first establishment core activities in updating and maintenance cannot be easily outsourced. In-house capacity to update and maintain spatial data is a necessity. Even if survey tasks are given to a private surveyor, in-house capacity is also important to control the private surveyors, to manage contracts, and to check outputs.

The speed for parcel surveying for a first systematic survey and for sporadic maintenance and updating surveying are not the same. The latter need more time, as the travel distance between the surveyed parcels is longer.

The Possibility to re-establish boundaries is very important in the maintaining process. During conflicts boundary marks can be destroyed, damaged or moved. If these marks cannot be trusted, surveyors are able to re-establish the boundary points with at least the same accuracy level of the original surveys.

The cost for parcel for establishment and the cost for upgrading and maintenance normally are different. Costs vary on the chosen survey methods and on the desired accuracy. In some cases one or two boundary point measurements can be enough to re-survey a parcel and the relative cost could be minimal. The important issue to consider during re-surveys is the participation and the understanding of the users. The method has to be simple enough to be understood by users. Re-surveys are needed for variety of reasons. One method cannot satisfy all needs. The way of working has to be open enough to accommodate combination of methods depending on the nature of the re-survey.

Orthophotos produced from aerial photos with a minimal ground resolution of 20 cm is identified to be the best method for the assessment of cadastral maps within the planned second level certification program in ANRS. The accuracy level is suitable for all holding types, the method is cost effective, the human power requirement is minimal, and lost boundary points can be re-established easily. The method is user friendly, the material and technologies to be apply is feasible also at local level. In short, mapping of parcels using orthophotos fulfils all stipulated selection criteria. Areas, requiring precision surveys, such as farm lands in modern irrigation schemes and high value parcels in peri-urban areas, need to be surveyed by RTK GNSS or with conventional total stations (see *Table 16*).

Table 16: Comparison of surveying methods

	RTK GPS	HH GPS	Orthophoto	HRSI
Accuracy in RMS ($\pm$ m)	Reference	4.1	0.61	1.50
Labor (par/sur/day)	30	30	40	40
Cost per parcel ($)	12.6	6.9	8	10
Re-establishment possibility	Excellent	Poor/No	Very Good	Good
Connection to a grid	Yes	Yes	Yes	Yes
Multipurpose	V good	good	Excellent	Excellent
Combination possibility	yes	yes	yes	yes
Existing resources	yes	yes	Going on	no

The method selected for the maintenance of the spatial contents in the land administration system is RTK GPS or total station ground surveys supported by the remote sensing tools.

6 Summary, Conclusions, and Recommendations

6.1 Identification of the Key Intervention Areas for Effective Development of Progressive Land Administration Systems Using ANRS Rural Land Administration as a Case

Four key intervention areas that are important to establish a progressive land administration system in the developing world were identified: Proper institutional set up, cadastral and registration proclamation, densified ground control points, and suitable cadastral survey methods. The systematic evaluation of the ANRS progressive land administration system and the description of the formal and the informal system using the legal cadastral domain model proved to be an effective tool to identify the key intervention areas and to lay a solid ground for knowledge exchange between systems.

6.1.1 Evaluation Progressive Land Administration Systems

The case study concludes that the framework developed for evaluating progressive land administration systems is an important tool. Including major land issues into the constitution is an advantage at policy level while developing new land administration systems for the African situation. The institutional mandate should be based on core land administration functions. Institutions need to consider customer satisfaction surveys as indicators for achievements.

The discussion on strengths, weaknesses, opportunities and threats on ANRS rural land administration system (SWOT analysis) considered issues related to policy and law, to management, to operational functions, to review processes, and to external factors.

It is recommended that the priority of activities should consider the prime objective. Measures to solve the major problems of tenure security have to be implemented based on their priority order. Cadastral land surveying with high accuracy and with high technical standards is not an obligatory pre-condition for tenure security. Prior implementations of legal reforms and a participatory way of working to address the urgent needs for sustainable projects are recommended. Embarking on costly cadastral system implementation before finding consent with landholders about the system to be implemented, can lead to costly surprises – especially in countries, where formal land administration systems are not common.

The study identifies, that large scale implementation of land use plans is constrained by technical complications and costs. It is believed that – in a first stage - land use plans can be prepared at parcel level based on contractual agreements with responsible landholders. The plans can be enforced by law or custom depending on the legal framework of a country. If the land surveying process will be completed, local level participatory land use plans can be upgraded to full-fledged detail plans by linking the parcel-based land use information to cadastral maps. Attempts to make corrective decisions on development activities without data input from land use plans can lead to subjective and unfair decisions. In the worst case it can be a threat for good land governance and can erode the public trust on land administration institutions.

The main strength of the ANRS land administration system is the implementation of the way of working that is a combination of legal and participatory processes. The major weakness identified during the study was the little attention given to land use planning and development control core functions. The planned second level certification program has to give due consideration for both legality and active participation of users.

6.1.2 Formal and Informal Property Right Systems

Both, formal and informal settings in ANRS can be described using the legal cadastral domain model (LCDM). The outlined investigation revealed the similarity between both systems. As main differences social advantages and sanctions in the informal setting were identified. The two seemingly different systems can be integrated at the grassroots level by incorporating the rules of the informal setting into the directives of the formal one. The directive is capable of addressing the required flexibility, and the peculiar nature of the rules of the informal setting in time and space.

The formalization process has special significance to countries, where the major insecurity problems are associated with state sponsored land redistribution. The formalization process or the introduction and implementation of the formal system have to be based on the basic rules of the informal setting. The integration of the two systems is only possible, if there is a proper understanding and description of each.

The reconciliation of legality in the formal setting with the legitimacy of the informal setting is important to get public acceptance during the implementation of the formal setting. The innovative informal structures cannot be simply dismissed as illegal activities. Eliminating or replacing customary tenure is often neither necessary nor desirable. The investigation of the existing system is important to capitalize on the available knowledge and to avoid the attempt of reinventing the same wheel.

The similarity of the formal and informal property right systems can be identified as one of the strong reasons for the successful accomplishments of the design and implementation of the formal land administration system in ANRS. For every introduction of a new property right system in Africa it is recommended to describe carefully the informal setting for the specific country and to try to incorporate as many rules as possible into the newly developed formal system.

6.2 Toolbox Guiding the Development of Methods for Identified Key Intervention Areas of Progressive Land Administration System

6.2.1 Identification of Tools for Organizing Effective Institutional Set up

The study identified important considerations to establish an effective institutional setting for progressive land administration systems. The main important considerations were:

- Legally supported and clear mandate for land administration institutions;
- Organizing all land administration functions in one institution (all in one type of institutional set up);
- Continuous capacity development in terms of human resource development, resources and working procedures;
- Effective communication and well established structure up to the grassroots level;
- Acknowledgment of customary rules and effective integration of informal setting;
- Organizing the institutional set up by making core land administration functions as core business processes;
- Increased private sector involvement.

6.2.2 Tools for the Development of Cadastral and Registration Proclamation

The objective of cadastral and registration projects is managing different interests on land. The study identified detailed provisions to be included in cadastral and registration proclamation in Ethiopia to manage interests on land. The list by its very nature cannot be exhaustive; therefore additions, modifications and even deletions with convincing reasons are highly appreciated. The interests on land were divided into three, namely the individual interests on land, the government management interests, and the government access interests on land. The private interest on land was sub-divided in to: rights, restrictions and responsibilities of landholders.

The aim of the development of this tool was detailing the land administration and cadastral principles to develop optional provisions to be considered in a cadastral and registration proclamation. The result of the study can be used to address the needs of large scale implementation of second level certification in Ethiopia.

The outlined optional provisions are based on the current experience and the needs reported in ANRS. The tool will be used as guiding document for the development of a federal framework law on cadaster and registration interests on land in Ethiopia. Additionally the regional states, such as the ANRS can use the tool to develop regional cadastral and registration proclamations based on the federal framework law.

The options can be modified or changed to address specific needs of each jurisdiction. So the tool might be a contribution to develop cadastral and registration laws for other developing countries with a similar socio-economic situation. It is recommended, that ANRS should enact cadastral and registration proclamation before launching the implementation of second level certification program.

6.2.3 Tools for Geodetic Control Points for Cadastral Surveying

Based on the result of the evaluation it can be stated that EMA network needs significant work on rehabilitation and upgrading. The density of ground control points in the area in general is very low to support the second level certification program. PPP (Precise Point Positioning) methods for the densification of geodetic points in the Amhara national region are recommended based on accuracy, time, and cost reasons.

Extensive densification of geodetic points is required for an effective implementation of second level certification program in the Amhara national region of Ethiopia. The current study concludes that the relative accuracy of PPP points is sufficient for the cadastral project. Previous disadvantages of PPP, such as slow convergence times, no user equipment supports real-time algorithms, no real-time satellite orbit and clock data streams, uncertain coordinate datum, are mainly solved by the recent developments. The optimum observation time needed for establishment of control points using static PPP method using double frequency GNSS was between 1 and 2 hours.

The study concludes as good example for grid point description the template used for the AM network. The descriptions have to include an overview location description using 1:50.000 scale topographic maps, pictorial descriptions, diagrams, sketches and textual descriptions.

New ground control points have to be established by marks drilled on stable bed rocks and – to guarantee an easy and quick assibilate – close to the existing road network. Protection and maintaining mechanisms have to be arranged in cooperation with local authorities. The measurements of control points have to be outlined by the regional authority, namely the BoEPLAU.

Information about ground control points has to be free for all users. Landholders nearby the control points have to be informed about the importance of the control points and they should get an especial responsibility to report any possible risk or damage on the control points.

6.2.4 Tools for Selection of Cadastral Surveying Methods

The developed world has relatively well developed spatial data infrastructure to facilitate sustainable development. There is a consensus on the contribution and importance of land administration systems for sustainable development in the developing world. Among others, the major cause of failure for progressive land administration systems in the developing world is lack of informed decision on establishment and maintenance of well-functioning cadastral systems. As long as the knowledge of how to bring about the best system in Sub-Saharan Africa is limited, studies will continue to look for solutions.

Within this study selection criteria to guide the second level certification program in ANRS were developed. Criteria considering aspect before launching the cadastral systems and technology selection decision criteria were outlined.

The case study conducted in the ANRS confirmed that the developed criteria can effectively guide the process for the selection of proper methods for well-functioning cadaster in progressive land administration systems. The status of objective and goal setting criteria in the Amhara national regional state is generally in very good status, but the results of the case study indicated that responsible authorities in the regional state have to take policy measures to improve the appropriateness of spatial data for the free market. Targets and indicators of achievement need to be detailed at the country level development plan documents. The implementing institutions have to be capacitated for outsourcing and administering technical contracts effectively.

The timing and need assessment decision criteria are used to check the proper timing for the commencement of cadastral surveys. The case of the Amhara national region proved that proper timing is an important success factor. The strategic decision to have primary and secondary levels of registration in ANRS was an excellent decision. Correct timing is expressed in terms of the readiness of implementing institutions, users, stakeholders and legal framework.

The study showed that the Amhara national regional state is in very good position concerning scheduling and readiness. However, the increase of skill level for control point establishment, the updating of textual data, and human power and budget allocation needs improvement. In addition to this, lack of definitions on data ownership, clear responsibilities for spatial data management and procedure for the use of spatial data by stakeholders' needs serious considerations.

It is recommended to enable stakeholder offices to use spatial data for their activities and to prepare rules concerning the ownership and responsibility of spatial data before launching the second level certification program.

The general evaluation result of the status of the way of working design decision criteria in the Amhara national regional state was very good. The issues that need some improvement are the lack of obligatory modalities for connecting spatial data infrastructure to the national grid, the lack of legal recognition for defined technical standards of different holding types, and the lack of proper rules to enforce spatial data updating and data migration to and from different data bases.

The ANRS land administration system is at the establishment stage for the second level certification program. Based on the field test conducted on existing remote sensing and ground survey tools, the use of orthophotos with the support of RTK GNSS measurements is recommended as proper method for the spatial assessment, the updating and the maintenance of land parcels.

7 References

Adal, Y. (2002). Review of Landholding Systems and Policies in Ethiopia under the different Regimes. Working paper No5/2002. Addis Ababa, Ethiopia: Ethiopian Economic Policy Research Institute.

Alemu, T. (2005). The Land Issue and Environment Change in Ethiopia. . Adis Abeba, Ethiopia: Adis Abeba University, Department of economics.

Alemu, B. Y. (2012). Expropriation, Valuation and Compensation Practice in Amhara National Regional State (ANRS) – The Case of Two Cities (Bahir-Dar and Gonder). Nordic Journal of Surveying and Real Estate Research 9:1, 30-58.

Ali, Z., Tuladhar, A., & Zevenbergen, J. (2010). Developing a Framework for Improving the Quality of a Deteriorated Land Administration System Based on an Exploratory Case Study in Pakistan. Nordic Journal of Surveying and Real Estate Research 7:1, 30-57.

Ambaye, D. (2013). Land Rights and Experoperation in Ethiopia, PhD thesis. Stockholm: Real Estate Planning and Land Law, Department of Real Estate and Construction Management, School of Architecture and the Built Environment, Royal Institute of Technology (KTH).

Andersson, B. (2005). Future Framework of Land Related Laws in Amhara Regional State: Study to Support Law Development. . Bahir Dar: SIDA-Amhara Rural Development Program, Ethiopia.

ANRS. (2000). Proclamation Issued to Determine the Administration and Use of the Rural Land in the Amhara National Region. Proclamation No.46/2000. Bahir Dar, Ethiopia: Zikre Hig.

ANRS. (2006). The revised Amhara National Regional State Rural Land Administration and Use Proclamation. No.133/2006. Bahir Dar, Ethiopia: Zikre Hig.

Aredo, D. (2003). Review of Theories on Land Tenure and Country Experiences. Adis Abeba, Ethiopia: The Ethiopian Economic Association/ Ethiopian Economic policy Research, Working paper series No. 4.

Ashenafi, Z. T., & Leader-Williams, N. (2005). Indiginous Common Property Resources. Human Ecology, 33(4), 539-563.

Augustinus, C. (2002). Comparative study of land administration systems –case study Uganda.

Augustinus. C. (2003). Comparative Analysis of Land Administration Systems: African Review, with special reference to Mozambique, Uganda, Namibia, Ghana, South Africa. Work undertaken for the Worldbank, funded by DFID.

AUSPOS. (2012). AUSPOS - Online GPS Processing Service. Geoscience Australia.

Barnes, G. (1990). A Comparative Evaluation Framework for Cadastre-Based Land Information Systems (CUS) in Developing Countries, University of Wisconsin Land Tenure Centre.

Bedada, T. B. (2010). Absolute geopotential height system for Ethiopia. Edinburgh: University of Edinburgh.

Bennett, R., Wallace, J. & Williamson, I. P. (2008). A Toolbox for mapping and managing new interests over land. Survey Review, 40(307), 43-53.

Bennett, R., & Rajabifard, A. (2009). The RRR Toolbox: a Conceptual Model for Improving Spatial Data Management in SDIs. SDI Convergence, 253.

Bezabih, M., Holden, S., & Mannberg, A. (2012). The Role of Land Certification in Reducing Gender Gaps in Productivity in Rural Ethiopia. CLTS Working Paper No. 1/2012.

Bezu, S. & Holden, S. (2013). Unbundling Land Administrative Reform: Demand for Second Stage Land Certification in Ethiopia , Ås, Norway: Centre for Land Tenure Studies, School of Economics and Business, Norwegian University of Life Sciences.

Blackwell, W. H. (1962). The adjustment of the Blue Nile Geodetic Control Project. Journal of Geophysical Research.

BNBSP (1961). Ethiopia Geodetic survey, (1957-1961). Ethiopia Geodetic survey report on horizontal and vertical surveys of the Blue Nile River basin by. Addis Abeba: U.S. Department of commerce coast and Geodetic survey for Ethiopia-United states Cooperative Program for Water Resources.

Bromley, D. (2008). Formalising property relations in the developing world: The wrong prescription for the wrong malady. Land Use Policy 26, 20–27.

Burns, T. (2007). Land Administration Reform: Indicators of Success and Future Challenges. 1818 H Street, NW, Washington, DC 20433: The International Bank for Reconstruction and Development / The World Bank.

Chambers, R. (1983). Rural Development: Putting the Last First. Harlow: Longman.

Chole, E. (1990). Agriculture and Surplus Extractions: The Ethiopian Experience. . In S. P. (Eds) (Ed.), Ethiopia Options for Rural Development. (pp. 89-99). London: Zed Books Ltd.

Corlazzoli, M. & Fernandez, O. (2004). SPOT 5 Cadastral validation project in Izabal,Guatemala,. s.l., FIG,Commission VII, Working Group VII/4..

Craig, Belle, A., Wahl, Jerry, L. (2003) Cadastral Survey Accuracy Standards. Journal of Survey and Land Information Science, Vol. 63, No. 2, 2003, pp 87-106.

CSA. (2007). Census report. Addis Abeba: Centeral statictics Authority, Ethiopia.

Cusworth, J., & Franks, T. (1993). Managing Projects in Developing Countries. Harlow: Longman.

Dale, P., & McLaughlin, J. (1999). Land administration. Oxford: Oxford University Press.

De Soto, H. (2000). The Mystery of Capital: Why Capital Triumphs in the West and Fails Everywhere Else. London: Bantam Press.

Deininger, K., & Jin, S. (2006). Tenure security and land-related investment: Evidence from Ethiopia. European Economic Review, 50, 1245–1277.

Deininger, K., Daniel, A., & Tilahun, A. (2011). Impacts of Land Certification on Tenure Security Investment, and Land Market Participation: Evidence from Ethiopia. Land Economics 87(2), 312-334.

Deininger, K., Daniel, A., Holden, S., Zevenbergen, J. (2008). Rural land certification in Ethiopia: Process, initial impact, and implications for other African countries. World Develpment, 36(10), 1786-1812.

Deininger, K., Jin, S., Adenew, B., Gebre-Selassie, S., Nega, B. (2003). Tenure Security and Land-Related Investment: Evidence from Ethiopia. , The World Bank: Policy Research Working Paper2991, Development Research Group

Diallo, A., & Thuillier, D. (2005). The success dimensions of international development projects, trust and communication: an African perspective. International Journal of Project Management 23 (3), 237–252.

Ebner, R., Featherstone W.E. (2008). How well can online GPS PPP post-processing services be used to establish geodetic survey control networks? Journal of Applied Geodesy 2(3): 149-157.

EEA. (2002). Land Tenure and Agricultural Development In Ethiopia. Adis Abeba: Ethiopian Economic Association / Ethiopian Economic Policy Research Institute.

El-Mowafy, A. (2011). Analysis of Web-Based GNSS Post-Processing Services for Static and Kinematic Positioning Using Short Data Spans. Survey Review Vol 43, No 323, 535-549.

EoE. (1960). Civil Code of the Empire of Ethiopia, Proclamation No. 165/1960. Addis Ababa, Ethiopia: Birhanena Selam Printing Press.

EOE. (1980). Proclamation no 193/1980 Proclamation for the establishment of Ethiopian mapping agency. Adis Abeba, Ethiopia: Berehanena Selam printing press.

Enemark, S., Clifford, K., Lemmen, C. & McLaren, R. (2014). Fit-For-Purpose Land Administration, Joint FIG / World Bank Publication, s.l.: The World Bank and the International Federation of Surveyors (FIG).

FAO. (2002). Land Tenure and Rural Development. Rome: Food and Agricultural Organization of the United Nations.

FDRE. (1995). The Constitution of the Federal Democratic Republic of Ethiopia, Proclamation No.1/1995. Addis Ababa, Ethiopia: Birhanena Selam Printing Press.

FDRE. (1997). Rural land administration proclamation of the Federal Government of Ethiopia: Federal Negarit Gazeta No. 89/1997. Addis Ababa: Berhanena Selam Printing Enterprise.

FDRE. (2005). A Proclamation to provide for the expropriation of land holdings for public purposes and payment of compensation. Federal Negarit Gazeta. Addis Ababa, Ethiopia: Berehanena Selam Press.

FDRE. (2005). Rural Land Administration and Land Use Proclamation No 456/2005. . Addis Ababa, Ethiopia: Birhanena Selam Printing Press.

Feder, G. (1999). Land Administration Reform: Economic Rationale and Social Considerations. Paper presented at the UN-FIG Conference on Land Tenure and Cadastral Infrastructures for Sustainable Development, Melbourne, Australia 25-27 October 1999.

FGCC (1984). Standards and Specifications for Geodetic Control Networks, Federal Geodetic Control Committee, Rockville, Maryland.

Fitzpatrick, D. (2006) .Evolution and Chaos in Property Rights Systems: The Third World Tragedy of Contested Access. The Yale Law Journal 115(5):996-1048.

Gakstatter, E. (2013) https://twitter.com/GPSGIS_Eric

Gedamu, A. (2009). Testing the Accuracy of Handheld GPS Receivers and Satellite Image for Land Registration, Stockholm, Sweden: Division of Geodesy Royal Institute of Technology (KTH).

Georgoulias, K., Papakostas, N., Makris, S., Chryssolouris, G. (2007). A toolbox approach for flexibility measurements in diverse environments. CIRP Annals-Manufacturing Technology, 56(1), 423-426.

Fradkin, K., and Doytsher, Y., (2002). Establishing an urban digital cadastre: analytical reconstruction of parcel boundaries. Computers, Environment and Urban Systems, Volume 26, Issue 5, September 2002, Pages 447-463.

Haile, Solomon. A (2005). Bridging the Land Rights Demarcation Gap in Ethiopia: Usefulness of High Resolution Satellite Image (HRSI) Data. PhD Thesis at the University of Natural Resources and Life Sciences (BOKU), Vienna.

Hanson, M. H., & Herz, R. S. (2011). A" Toolbox Approach" for Developing Thoughtfully Structured, Creative Art Experiences. Art Education, 64(1), 33-38.

Hanstad, T. (1997). Designing Land Registration Systems for Developing Countries. Am. U. Int'l L. Rev., 13, 647.

Hodgson, S. (2004). Land and Water-The Right Interface. FAO, Rome: Food and Agriculture Organization of the United Nations, LSP Working paper 10.

Holden, S. T., Deininger, K., & Ghebru, H. (2009). Impacts of Low-cost Land Certification on Investment and Productivity. American Journal of Agricultural Economics 91 (2), 359-373.

Holden, S. T., Deininger, K., & Ghebru, H. (2011). Tenure Insecurity, Gender, Low-cost Land Certification and Land Rental Market Participation. Journal of Development Studies 47(1), 31-47.

Holden, S., & Ghebru, H. (2013). Welfare Impacts of Land Certification in Tigray, Ethiopia. In S. Holden, K. Otsuka, & K. Deininger, Land Tenure Reforms in Asia and Africa: Impacts on Poverty and Natural Resource Management (p. Chapter 6). Palgrave Macmillan.

IAAO. (2004). Standard on Manual Cadastral Maps and Parcel Identifiers. 130 East Randolph, Suite 850, Chicago, IL 60601-6217: International Association of Assessing Officers.

ISO. (2012). ISO 19152:2012, Geographic Information – Land Administration Domain Model. Edition 1, 118 p., Geneva, Switzerland.

ISS. (2013). Comparative evaluation of high resolution satellite imagery against total station for cadastral surveying and mapping of rural lands, Addis Abeba: ELAP.

Jing, Y., Bennett, R., Zevenbergen, J. (2013). Up-to-dateness in land administration: setting the record straight. In: Environment for Sustainability, Paris.

Kalbro, T. (1996). Aspects of Permit Procedures for changes in Land Use. In Land Law in Action, A collection of contributions by Participants in the Seminar on the Theme land 87 Reform including Land Legislation and Land Registration. Stockholm, Sweden: The Swedish Ministry of Foreign Affairs, Division for the Central and Eastern Europe and The Royal Institute of Technology.

Kiril, F. & Yerach, D. (2002). Establishing an urban digital cadastre: analytical reconstruction of parcel boundaries Computers, Environment and Urban Systems vol 26, 447-463.

Kronman, A. (1985). Contract law and the state of nature. Journal of Law, Economics and Organization, 5–32.

Lane, M. (2001). Indigenous Land and Community Security: A (Radical) Planning Agenda. Land Tenure Center University of Wisconsin-Madison, Working paper 45.

Larsson, G. (1991). Land registration and cadastral systems, tools for land information and managment. Harlow: Long man scintific and technical.

Lemmen, C. (2012). A Domain Model for Land Administration, PhD thesis, Delft, the Netherlands: ITC, Sieca Repro BV.

Lemmen, C., Augustinus, C., Haile, S., Oosterom, P. (2009). The Social Tenure Domain Model – A Pro-Poor Land Rights Recording System. GIM.

Lemmen, C. & van Oosterom, P. (2006). Version 1.0 of the FIG Core Cadastral Domain Model. XXIII FIG Congress – shaping the Change, Munich, Germany, October 8-13, 2006

Maathuis, B., Mannaerts, C., & Retsios, B. (2008). The ITC GEONETCast-toolbox approach for less developed countries. Proceedings of Commission VII, 37, 1301-1306.

Mathison, S. (1988). Why triangulate? (H. Stuckensmidt, E. Stubkjaer, & C. Schlieder, Eds.) Educational Researcher, 17(2), 13-17.

Mattsson, H. (2003). Aspects of Real Property Rights and their Alteration. . In The Ontology and Modelling of Real Estate Transactions (pp. 23-34). In Stuckenshmidt, Stubkjær and Schlieder (Ed.) ISBN 0-7546-32873.

Mayoux, L. (2006). Quantitative, Qualitative or Participatory? Which Method, for What and When? Doing Development Research. London: V. Desai and R. B. Potter London Sage Publications Ltd.

Mesfin, W.-M. (1991). Suffering under God's Environment, a vertical study of the predicament of peasants in North-Central Ethiopia. Bern, Switzerland: The African Mountain association and Geographica Bernensia, Institute of Geography, University of Bern.

Miskas, A. & Molnar, A. (2009). Establishing a Reference Network in Parts of Amhara Region, Ethiopia Using Geodetic GPS Equipment. Stockholm, Sweden: Division of Geodesy Royal Institute of Technology (KTH)

Miskas, A. & Molnar, A. (2010). The AM network, a bgeodetic referance network covering the Amhara region in Ethiopia. Bahir Dar: BoEPLAU.

Mitchell, D., M. Clarke and J. Baxter (2008). Evaluating Land Administration Projects in Developing Countries. Land Use Policy, 25:464-473.

MOFED. (2010). The Federal Democratic Republic of Ethiopia Growth and Transformation Plan (GTP) 2010/11-1014/15. . Addis Ababa.: MoFED.

Moreno, B., Radicella, S., De Lacy, M. C., Herraiz, M., G. & Rodriguez-Caderot (2011). On the effects of the ionospheric disturbances on precise point positioning at equatorial latitudes. GPS solutions, 381-390.

Mugnier, C. (2003). Grids and Datums, the Federal Democratic Republic of Ethiopia. Loulslana: the American Society for Photogrammetry and Remote Sensing.

NASA. (2012). California Institute of Technology. The Automatic Precise Positioning Service

Nega, B., Adenew, B. & Gebre Sellasie, B. (2003). Current Land Policy Issues in Ethiopia. , Addis Ababa, Ethiopia: Ethiopian Economic Policy Research Institute.

North, D. C. (1993). Nobelpriz.org. Retrieved from nobel_prizes: http://nobelprize.org/nobel_prizes/economics/laureates/1993/north-lectu

NRCANGSD. (2012). Natural Resources Canada. Precise Point Positioning.

Obeng-Odoom F. (2012). Land reforms in Africa: Theory, Practice, and Outcome, Habitat International, 36, pp. 161 – 170.

Ostrom, E. (1998). Efficiency, sustainability and access under alternative property regimes. . Santiago Chilli: IMU/WIDER project.

Paasch, J. (2008). Standardization within the Legal Domain: A Terminological Approach. In: Euras Yearbook of Standardization, vol. 6. s.l.: online publication, pp. 105-130.

Paasch, J. (2005). Legal Cadastral Domain Model. An object-orientated approach. Nordic Journal of Surveying and Real Estate Research, 2(1), pp. 117-136.

Paasch, J. (2011). Classification of Real Property Rights. A Comparative Study of Real Property Rights in Germany, Ireland, the Netherlands and Sweden, Stokholm: Royal Institute of Technology, KTH.

Paasch, J. (2012). Standardization of Real Property Rights and Public Regulations. The Legal Cadastral Domain Model, Stockholm, Sweden: KTH Architecture and the Built Environment Real Estate Planning and Land.

Paasch, J. , Oosterom, P. van, Paulsson, J. & Lemmen, C. (2013a). Specialization of the Land Administration Domain Model (LADM) – An Option for Expanding the Legal Profiles. Abuja, Nigeria, FIG.

Paasch, J., Oosterom, P. van, Lemmen, C. & Paulsson, J. (2013b). Specialization of the Land Administration Domain Model (LADM) - Modeling of Non-formal RRR. Kuala Lumpur, Malaysia, FIG.

Palm, L. (2006) Comparison of Total Station/Advanced GPS Survey and High Resolution Satellite Imagers. The National Conference on Standardization of Rural Land Registration and Cadastral Survey Methodologies, Addis Abeba. ELTAP.

PMGE. (1975). Proclamation for Public Ownership of Rural Land in Ethiopia, No 31/1975. Addis Ababa, Ethiopia: Birhanena Selam Printing Press.

Popovas, D. (2001). Adjustment of the Lithuanian GPS network using GAMIT. Master Theses. Aalborg University, Institute of Development and Planning.

Rahmato, D. (2005). From Heterogeneity to Homogeneity: Agrarian Class Structure in Ethiopia since the 1950s. Adis Abeba,Ethiopia: Forum for Social Studies.

Rahmato, D. (2011). Land to Investors: Large-Scale Land Transfers in Ethiopia. Forum for social studies.

Rizos C., Janssen V., Roberts C., Grinter T. (2012). Precise Point Positioning: Is the Era of differential GNSS positioning drawing to an end?, TS09B, FIG Working Week 2012, Knowing to manage the territory, protect the environment, evaluate the culture heritage, Rome, Italy, 6-10, May 2012.

Rui, F. & Salah, M. (2013). Permanent Stations Guidelines: Requirements and Installations (final draft). Nirobi: AFREE.

Seredovich, V., Irughe, R. & Ehigiator, M. O. (2012). PPP Application for estimation of precise point coordinates – case study of a reference station in Nigeria. Knowing to manage the territory, protect the environment, evaluate the cultural heritage, 2012 Rome, Italy. FIG Working Week.

Sheehan, J. (2001). Conceptualizing Native Title as an Analogues Property Right with in an Anglo-Australian Land Law Paradigm. Brisbane, Queens land, 400 Australia: Queens land University of technology.

Shibeshi, G., Fuchs, H. & Mansberger, R. (2013). Participatory and Pro-Poor Land Administration System of the Amhara National Regional State of Ethiopia: Lessons and Evaluation of the Current Status. Washington DC, The World Bank.

Shibeshi, G., Fuchs, H. & Mansberger, R. (2014). Toolbox for the Development of Cadastral and Registration Proclamation for Second Level Certification Program in Ethiopia. International Journal of Sciences: Basic and Applied Research (IJSBAR) ISSN 2307-4531, 13(1), pp. 244-259.

Simon, D. (2006). Your Questions Answered? Conducting Questinnaire Surveys Doing Development Research. London: V. Desai and R. B. Potter Sage Publication Ltd.

Smajgl, A., & Larson, S. (2007). Institutional Dynamics and Natural Resource Management. . In A. Smajgl, & S. Larson, Sustainable Resource Use: Institutional Dynamics and Economics. London: Earthscan.

Steudler, D., Rajabifard, A., & Williamson, I. P. (2004). Evaluation of land administration systems. Journal of Land Use Policy (21), 371 – 380.

Steudler D., Williamson IP. (2005). Evaluation of National Land Administration System in Switzerland. Survey Review, 38(298): 317–330.

Steudler, D., Williamson, I. P., Rajabifard, A., Enemark, S. (2012). Cadastral template [Online] Available at: http://www.cadastraltemplate.org/

Sultan, M. (2011). Geospatial capacity building and knowledg transfer. Addis Abeba: Ethiopian Maping Agency (EMA).

Ubink, J. M., & Quan, J. F. (2008). How to combine tradition and modernity? Regulating customary land management in Ghana. Land Use Policy, 25(2), 198-213.

UNECE. (1996). Land administration guidelines with special reference to countries in transition. Retrieved May 18, 2012, from http://www.unece.org/hlm/wpla/publications/laguidelines

UNECE. (2001). Land (real estate) mass valuation systems for taxation purposes in Europe. Geneva, Switzerland: Working Party on Land Administration.

Van Den Brink, R. (2002). Land Policy and Land Reform in Sub Sahara Africa Consensuses Confusion and Controversy. Presentation to the symposium Land redistribution in Southern Africa. Pretoria, South Africa.

Van der Molen, P. (2002). The dynamic aspect of land administration: an often-forgotten component in system design. Computers, Environment and Urban Systems 26, 361–381.

Williamson, I. (2000). Best Practices for Land Administration Systems in Developing Countries. International Conference on Land Policy Reform. Jakarta: LAP-C Project, Support for Long Term Development of Land Management Policies.

Williamson, I. (1981). The cadastral survey requirements of developing countries in the pacific region. With particular reference to fiji. Sydney, Australia, University of New South Wales.

WILLIAMSON, I. (1983). Cadastral Survey Techniques in Developing Countries with Particular Reference to Thailand. The Australian Surveyor vol 31, 496-512.

Williamson, I., Enemark, S., Wallace, J., Rajabifard, A. (2010). Land Administration for Sustainable Development. New York: ESRI Press.

Williamson, I. & Fourie, C. (1998). Using the Case Study Methodology for Cadastral Reform. Geomatica 52(3), 283–295.

Williamson, I. & Ting, L. (2001). Land Administration and Cadastral Trends-a framework for Re-engineering. Computers Environment and Urban Systems, Issue 25, pp. 339-366.

Woldeyes, Z. H. & Harris, D. (2014). Harnessing Synergies for Implementation and Monitoring Impact Rural Land Registration in Ethiopia Increased Transparency for 26,000,000 Land Holders. Washington DC, Land and Poverty Conference, the World Bank.

Yersaw, Alemu, B. (2012). Expropriation, Valuation and Compensation Practice in Amhara National Regional State (ANRS) – The Case of Two Cities (Bahir-Dar and Gonder). Nordic Journal of Surveying and Real Estate Research 9:1, 30-58.

Zevenbergen, J. (2002). Systems of land registration. Aspects and effects. PhD thesis, Delft, The Netherlands: Delft University of Technology.

Zevenbergen, J. (2004). A systems approach to Land Registration and Cadastre. Nordic Journal of Surveying and Real Estate Research.

8 Index of tables

Table 1: Standard of geodetic control points (source FGCC, 1984) 24

Table 2: Summery questions rated from 5 (excellent) to 1 (poor) based on respondent's experience. (Number of samples: 15 groups with total attendants of 70) 35

Table 3: Summery rating of policy and cross cutting issues from questioner survey (number of samples 118 offices) 38

Table 4: Summery rating for tenure core function (Number of total respondents=118) 43

Table 5: ISLA encoded data 44

Table 6: Summery table for the rating of value core function 47

Table 7: Summery of the replies of major stakeholder institutions 50

Table 8: Summery of SWOT analysis.. 55
Table 9: Ratings given by land administration professionals on the status of different land administration activities in the ANRS.. 57
Table 10: Correlations of AM and PPP measurements.. 78
Table 11: Availability and status of EMA points... 81
Table 12: Comparison of EMA points and their PPP calculated equivalent 86
Table 13: Correlations of EMA and PPP measurements 87
Table 14: Correlations between different time length static measurement results 88
Table 15: Cost and working hour calculated by ELAP pilot project (Source: ELAP report).. 99
Table 16: Comparison of surveying methods 100

9 Table of figures

Figure 1: Planned and existing ground control point distribution in the study area (source; based on Sultan, 2012) 25

Figure 2: Location of ANRS 28

Figure 3: Research design 29

Figure 4: Elements of Evaluation framework 34

Figure 5: Proposed organogram 41

Figure 6: Share % of investment land area for Woreda, (source: BoEPLAUA 2010) 46

Figure 7: Human power and budget increase in sample Woredas (district level) 49

Figure 8: Hierarchy of federal and regional laws in Ethiopia (Source: Anderson, 2005) 59

Figure 9: Modified legal cadastral domain model (LCDM) representing ANRS formal legal system (based on Paasch, 2012) 61

Figure 10: Legal Cadastral Domain Model (LCDM) representing the informal setting in ANRS (based on Paasch, 2012). 66

Figure 11: Point distribution of AM network 75

Figure 12: Picture of an GCP in the AM network (Source: Miskas & Molnar, 2010) 76

Figure 13: Template for an AM grid point (Source: Miskas & Molnar, 2010) 77

Figure 14: Picture showing grid point *(AM1-17 Adis Zemen)* 77

Figure 15: Existing and planned distribution of ground control points by EMA (source; Sultan, 2012) 79

Figure 16: Picture showing Adet EMA point 83

Figure 17: Example for older EMA point descriptions 83

Figure 18: Example for newer EMA point descriptions 85

Figure 19: Root mean square error for different observation times [in ± m] 88

Figure 20: User participation during land certification process 94

Figure 21: Examples of boundary types (furrow) 96

Figure 22: examples of boundary marks by stone lines and grass strip 96

Figure 23: Cadastral index maps produced using HRSI and distribution of sample points 97

Figure 24: Cadastral index maps produced using orthophoto and distribution of sample points 98

10 Appendix

10.1 Discussion Points with the Land Administration Committee Members

The objective of the survey is to get information about the status of land administration system in ANRS. The involvement and contribution of the land administration committee members is the special focus.

The hypothesis of the study is the efficiency of land administration system can be increased by establishing geo-referenced cadastral system that is connected to the national grid.

1. What are the major rights and restrictions:-
   - On individual possessions
   - On communal grazing lands
   - On communal forest lands
2. Are these rights and restrictions attached to the land or to the land holder?
3. How many landholders are living in your Kebele?
4. What is the average size of land holding in your Kebele?
5. What is the average size and number of parcels in one landholding?
6. Do you believe that issuance of secondary certificates with coordinated maps is necessary for your area?
7. How many cases do you see per week? (estimated average)
8. What are the major causes of conflict in your area?
9. Are boundary conflicts common? How much boundary shift is tolerated?
   - On individual holdings?
   - On communal holdings?
   - On irrigation areas?
10. What is the perception of the society about the role of your committee (as legal or customary)?
11. Have you ever experienced cases not covered by the land law, if yes what are some of the examples and how do you solve them?
12. Is the role of arbitration committee supportive to your duties?
13. What are the effects of land registration on rental market and share cropping?
14. Is it possible to incorporate the traditional land administration rules in your formal duty? If no what is the reason?
15. What are you incentives to be involved in the land administration implementation program? How much time do you spend to accomplish your role?
16. Do you get any support from the society? If yes, in what form?
17. Do you have your own office and archive for documents?
18. Do you get enough support from Kebele land administration expert?
19. Can you suggest some corrections to be made on land administration?

20. Any other points to comment?

Summery questions

Rate the questions from 5 (excellent) to 1 (poor) based on your experience

Sr no.	Question	Rate (5-1)
1	Are you satisfied with the participation of the landholders in your area?	
2	How important is land administration system for reduction of land related conflicts?	
3	Are landholders in your area willing to cover all costs of efficient land administration services?	
4	How impermanent is the spatial data set for land administration in your area?	

10.2 Discussion Points with Selected Woredas, Zones and Regional Land Administration Professionals

The objective of the survey is to get information about the status of land administration system in ANRS. The evaluation of the status of the current system and recommendations for suitable spatial framework is the special focus.

The hypothesis of the study is the efficiency of land administration system can be increased by establishing geo-referenced cadastral system that is connected to the national grid.

1. Discuss the mission, Vision and strategy for land administration?
2. What are the major objectives of land administration system in your office? Do you think they are correct and connected to mission, vision and strategy?
3. Is there a capacity to implement second level certification in your area?

    3.1 In terms of staff (skilled surveyors)

    3.2 In terms of equipment and other resources

    3.3 In terms of geodetic control points

    3.4 In terms of working procedures

4. What are the positional and relative accuracy needs for:

    4.1 Private holdings

    4.2 Communal holdings

    4.3 State holdings and investment lands

    4.4 Kebele boundary

    4.5 Location of houses and other important features

    4.6 Irrigation and height value lands

5. Can you discuss the major achievements of implementation of land administration system in your area?

    5.1 Achievements on land registration (Coverage of issuance of primary book of holdings)

    5.2 Achievements on cadastral surveying

- 5.3 Achievements in preparation of land use plans (strategic and local level)
- 5.4 Achievements in development control
- 5.5 Achievements on land valuation and compensation
- 5.6 Achievements in policy and legal system development and implementation
- 5.7 Achievements in computerization of documents
- 5.8 Achievements in public awareness and education
- 5.9 Achievements in capacity building
    - 5.9.1 How many professional were deployed in2003, 2008, and now in what field of study and educational level
    - 5.9.2 Budget assigned for LA in the respective years
    - 5.9.3 Status of Woreda and Kebele office and equipment
    - 5.9.4 Status of in-service and on job trainings
6. What are the major strengths of the system?
    - 6.1 Related to cadastral surveying and land registration
    - 6.2 Land use planning
    - 6.3 Land valuation
    - 6.4 Development control (Land use change control)
    - 6.5 Organizational setting
    - 6.6 Gender issues
7. What are the major weak points of the system?
    - 7.1 Related to cadastral surveying and land registration
    - 7.2 Land use planning
    - 7.3 Land valuation
    - 7.4 Development control (Land use change control)
    - 7.5 Organizational setting
    - 7.6 Gender issues
8. What are the opportunities?
    - 8.1 Political and legal environment
    - 8.2 Economic
    - 8.3 Social
    - 8.4 Technological
9. What are the major threats?
    - 9.1 Political and legal environment
    - 9.2 Economic
    - 9.3 Social
    - 9.4 Technological

10. What major changes do you propose in short and long terms?

11. Any other comments

Summery questions

Rate the questions from 5 (excellent) to 1 (poor) based on your experience.

Sr no.	Question	Rate (5-1)
1	Do you evaluate the land administration system in your area as a successful system?	
2	How important is land administration system for reduction of land related conflicts?	
3	Are landholders in your area willing to recover all costs of efficient land administration services?	
4	How impermanent is the spatial data set for land administration in your area?	

10.3 Questioner for Individual Farmers

The objective of the survey is to get information about the status of land administration system in ANRS. The understanding of landholders' expectations and satisfactions is the special focus of this questioner.

The hypothesis of the study is the efficiency of land administration system can be increased by establishing geo-referenced cadastral system that is connected to the national grid.

1. Personal details
    a. Name
    b. Sex
    c. Age class
    d. No of family members
    e. Marital status
2. Are you the landholder in the Kebele?
3. How many parcels do you hold?
4. How much is the area of each?
5. Do you think the amount of tax you pay is fair?
6. Do you earn additional income other than farming? If yes, what are the major sources?
7. Do you rent or rented land, if yes, how much and in what terms? If yes, is the contract officially registered?
8. What are the major causes for conflict in your area?
9. Do you ever come across land related conflicts, if yes how do you resolve them; are you satisfied with the solution?

10. Do you come across boundary conflicts? If yes for how much distance shift?
11. What are the major benefits you expect from second level of certification with coordinated parcel maps?
12. Do you believe that your holdings are secured?
    a. General (legal) security?
    b. Boundary conflict and local insecurity?
    c. Expropriation?
13. What other benefits do you get from land administration authorities?
14. What are the major causes of eviction from land?
15. Do you know anyone in your area who is evicted from his land by land administration decision?
16. Do you actively participate in land administration meetings, public hearings, land administration conference etc.? Why? Why not?
17. Are you satisfied with the service delivery of land administration authority?
    a. Is fee fair?
    b. Is customer treatment good?
    c. Are rules and regulations based on your interests?
    d. Are you satisfied with accuracy of spatial data?
    e. Is the land register updated regularly?
18. Do you up date changes due to land transaction on the register?
19. Is the fee for updating fair? If not, why not?
20. What improvements do you made on your land after your land is registered?
21. What are your rights and responsibilities on the communal lands?
22. Are you satisfied the way communal lands are managed in your area?
23. What are the major services you expect from land administration authorities?

Summery questions

Rate the questions from 5 (excellent) to 1 (poor) based on your experience.

Sr no.	Question	Rate (5-1)
1	Do you fully participate in land administration activities in your area?	
2	Are you satisfied with the current land administration services?	
3	Are you willing to cover all costs of efficient land administration services?	
4	Are coordinated parcel maps and second level of book of holding important to you?	

10.4 Questioner to be Filled by Major Stakeholders of Land Administration System in ANRS of Ethiopia

The objective of the survey is to get information about the status of land administration system in Amhara region. The involvement and contribution of the major stakeholders is the special focus.

The hypothesis of the study is the efficiency of land administration system can be increased by establishing geo-referenced cadastral system that is connected to the national grid.

1. Does your office have access to land administration data set? If yes, which data sets? And in what form?
   a. Analogue?
   b. Digital?
   c. Conventional post?
   d. Personal visits?
2. If yes on what basis?
   a. For free?
   b. On payment basis?
   c. By asking official cooperation?
   d. Other (specify)?
3. Are you satisfied with services you get from the land sector? Why?
4. Do you believe that land administration system has any contribution for accomplishments of your tasks? If yes, please list the major contributions?
5. What are the functions in your office that are directly or indirectly dependent on or significantly affected by land administration activities?
6. Do you come across any significant change in your system because of the influence from land administration and related activities?
7. Do land administration activities create any threat to any of your office interests? If yes what are these threats?
8. What are the contributions of spatial data sets for accomplishments of the responsibilities of your organization?
9. Do you have any quality requirements for spatial data sets in terms of:
   a. Accuracy
   b. Details
   c. Level of update
   d. Coverage
10. How much can your organization pay for land administration data sets?
    a. Spatial data sets
    b. Textual data sets

11. What contributions do you expect from land sector for efficiency and effectiveness of your work?
12. What type of change/ improvement do you suggest in the land sector? Why do you suggest these changes?
13. What can your office contribute for better land administration system in the country?
14. Do you have any comments on land administration system in the country?

Summery questions

Rate the questions from 5 (excellent) to 1 (poor) based on your experience.

Sr no.	Question	Rate (5-1)
1	Are you satisfied with services from land sector?	
2	Do you believe spatial data set is important for your activities?	
3	Are you willing to cover all costs of efficient land administration services?	

10.5 Questioner to be Filled by Land Administration Offices

Name:
Office:
This questionnaire is designed to collect background information about the status of the implementation of Land administration system in ANRS. The background information is collected as part of PhD study on Cadastral Procedure and Spatial Framework for the Development of Efficient Land Administration System in Ethiopia. The main objective of the study is to investigate the procedures, technologies and standards that could be used to satisfy Ethiopia's parcel mapping needs for improving Ethiopia's land records system. Make suggestions on technologies that are cost effective, pro poor, speedy and can easily be applied in Ethiopia. In general the objective of the study is to contribute for the development of efficient land administration system that can address the possible gaps in the existing system. This questioner deals with five major parts namely; general issues, land value issues, land use planning issues, and land development issues. The strength, weakness, opportunity and threat of the system in relation to the core functions of land administration system is the focus of this inquiry. The discussion points identified will be rated 1 to 5 based on the experience of the respondent in his/her jurisdiction.

		Your Score (5,4,3,2,1)	
	JUST AS A STARTER		
S1	How important is tenure security for economic development?		(5=very important, 1=not

S2	How important is the contribution of land administration system for environmental protection and natural resource conservation?		(5=very important, 1=not important)
S3	How important are clear cadastral procedures and coordinated cadastral maps for implementation of modern land administration system in your Woreda?		(5=very important, 1=not important)
S4	Is land administration data important for efficient tax collection in your Woreda		(5=very important, 1=not important)

A	General Issues		
A1	**Policy formulation & monitoring**		
A1.1	How important is a Land Administration strategy for a sustainable development of your Woreda?		(5=very important, 1=not needed)
A1.2	Do you feel that you were involved in the formulation of current land administration strategy to date?		(5=strongly, 1=not involved)
A1.3	Do you feel that this is necessary to be involved in the formulation of land administration strategy in the future?		(5=strong involvement, 1=less involvement)
A1.4	Do you think that policy revision and updating is conducted appropriately		(5=Very appropriate, 1= not conducted at all)
A1.5	Do you agree in making land issues of constitutional category?		(5=strongly agree, 1=do not agree)

A2	**Legal**		
A2.1	Do you think gaps and inconsistencies in the legal system are causing big challenge in your daily work?		5=big challenge, 1=no challenge
A2.2	Do you think that the legal framework for land administration system should be radically changed to improve effectiveness of your daily work?		(5=strongly, 1=less)
A2.3	How important is law in cadastral procedures		(5= very important, 1=less important)
A2.4	Do you have sufficient support from legal specialists in respect of legal issues?		(5=sufficient, 1=not sufficient)
A3	**Computerization and IT Systems**		
A3.1	How would you rate your level of knowledge on the ISLA (Information System for Land Administration)?		(5=enough, 1=too little)
A3.2	Do you use ISLA for land registration in your office?		(5=every day, 1=never)
A3.3	Is the current IT-Infrastructure sufficient for you?		(5=sufficient, 1=not sufficient)
A3.4	Is the current internet connectivity, capacity and speed enough for data exchange		(5=enough, 1=not enough)
A3.5	Is the current bandwidth (speed) of Woredanet sufficient and accessible for you?		(5=good, 1=bad)
A3.6	How secure is data in your system?		(5=well secured, 1=note secured)
A3.7	How reliable are your current system backup procedures ?		(5=good, 1=bad)
A4	**Research & Development**		
A4.1	How much do you benefit from current research and development from?		(5=much, 1=low)
A4.2	Do you feel that you are involved in advising the current research and development?		(5=much, 1=low)
A5	**Capacity building**		
A5.1	Is the human resource development program in your woreda sufficient?		(5=sufficient, 1=not sufficient)
A5.2	Are working procedures for land administration sufficient?		(5=sufficient, 1=not sufficient)

A5.3	Do you have enough field equipment for land administration tasks in your Woreda?		(5=sufficient, 1=not sufficient)
A5.4	Is the office space enough and suitable to undertake land administration tasks?		(5=sufficient, 1=not sufficient)
A5.5	Do you have enough transportation facilities to perform your tasks?		(5=sufficient, 1=not sufficient)
A5.6	Are archives to store land related data secure and safe?		(5=safe, 1=not safe)
A5.7	Do you have enough office consumables for land administration tasks in your Woreda?		(5=enough, 1=not enough)

A6	**Public information and awareness**		
A6.1	Is the information about land administration system enough for landholders in your area?		(5=enough, 1=not enough)
A6.2	Do mass medias in your area give enough coverage for land administration education?		(5=enough, 1=not enough)
A6.3	Are the landholders in your Woreda informed about their rights and obligations?		(5=well informed, 1=not informed)
A7	**Finance and cost recovery**		
A7.1	Are fees for land administration services fair?		(5=fair, 1=not fair)
A7.2	Are the payments by investors for lease contracts fair?		(5=Over valued, 1=under valued)
A7.3	Is the cost recovery system in place effective?		(5=very effective, 1=not effective)
A7.4	Is rural taxation dependent on land administration data?		(5=fully dependent, 1=not dependent)
A7.5	Can land administration generate enough finance to cover its costs?		(5=fully cover, 1=no contribution)
A8	**Research and development**		
A8.1	Are you getting enough support from current land administration research?		(5=enough, 1=no support at all)
A8.2	Are you involved in setting research agenda for land administration?		(5=involved at all stages, 1=not involved)
A8.3	Is the current land administration research focused on solving your problems in the field?		(5=fully problem solving, 1=no research at all)
A8.4	Is the land administration research conducted in your area used participatory approaches?		(5=fully participatory, 1=no research at all)

A8.5	Do you think professional debates in the country are focusing on proper issues?		(5=fully, 1=not at all)
A9	**Gender issues**		
A9.1	Are women's' land rights fully protected in your Woreda?		(5=fully protected, 1=not protected)
A9.2	Do women have equal access to land rights in your woreda?		(5=equal access, 1=no access)
A9.3	Are women involved in land administration decisions?		(5=fully involved, 1=not involved)

Please use this space to provide any comments on this topic that you feel may be helpful:

B	Land Tenure		
B1	Adjudication		
B1.1	How important was the past land redistribution data for current adjudication in your woreda?		(5 important, 1= not available)
B1.2	Do people participate during adjudication process in your woreda?		(5=a lot, 1=none)
B1.3	Are land holders satisfied with the content and shape of book of holdings and register book?		(5=satisfied, 1=not satisfied)
B1.4	Is systematic and compulsory registration system difficult to implement in your woreda?		(5=very difficult, 1=note difficult)
B1.5	Are all fixtures on landholdings registered during adjudication?		(5=all registered, 1=none registered)
B1.6	How important are public hearings for transparences and minimizing corruption?		(5=a lot, 1=none)
B2	Transfer		
B2.1	Do people report and register land transfer by inheritance?		(5=Report all,1=report none)
B2.2	Do landholders report and register land transfer by gift?		(5=Report all,1=report none)
B2.3	Are fees for land transactions acceptable by landholders in your Woreda?		(5=acceptable, 1=not acceptable)
B2.4	How common is land transfer for consolidation in your area?		(5=very common, 1=none)
B2.5	How much is the involvement of the spouse during land transaction		(5=no transaction without consent, 1=not involved)
B3	Land lease and rental contracts		
B3.1	Are all lease agreements with investors registered in your office?		(5=all, 1=none)
B3.2	Are all rental agreements with the duration of more than 3 years registered in your office?		(5=all, 1=none)
B3.3	Do all investors comply to the contractual agreements??		(5=all, 1=none)
B3.4	Is the fee for registration of contracts acceptable by the public?		(5=acceptable, 1=not

			acceptable)
B4	**Updating**		
B4.1	Is the land administration data contained in your system up-to-date?		(5=up-to-date, 1=obsolete)
B4.2	Is the land administration data contained in your system accurate?		(5=accurate, 1=not accurate)
B4.3	Are people willing to update land administration data in time?		(5=willing, 1=not willing)
B4.4	Is the procedure for updating simple?		(5=very simple, 1=complicated)
B5	**Unique parcel identifiers**		
B5.1	Are alpha numeric unique parcel identifiers easy to encode and use?		(5=very easy, 1=complicated)
B5.2	Is unique identified at the landholding level important for daily use?		(5=very important, 1=not important)
B5.3	Are alpha numeric unique parcel identifiers complicated for computerization?		(5=very easy, 1=complicated)
B6	**Boundary monuments**		
B6.1	Do you mark permanent boundary marks during adjudication?		(5=always, 1=never)
B6.2	Are boundary marks visible on satellite images and aerial photos?		(5=clearly visible, 1=not visible)
B6.3	How often are boundary marks are destroyed or moved?		(5=always, 1=never)
B.7	**Ground control points**		
B7.1	Are ground control points accurate enough for cadastral surveying purposes?		(5=always, 1=never)
B7.2	Are ground control points accessible for your daily use?		(5=always, 1=never)
B7.3	Can you easily locate ground control points based on their site descriptions?		(5=very easily, 1=not at all)
B.8	**Cadastral survey**		
B8.1	Are techniques and technologies used for cadastral surveying suitable for your woreda?		(5=very suitable, 1=not at all)
B8.2	Do you perform cadastral surveys in your Woreda?		(5=every day, 1=not at all)
B8.3	Is the cadastral data in your area connected to land registration data?		(5=fully integrated, 1=not at all)
B8.4	Is private sector involved in cadastral surveying in your Woreda?		(5=very frequently,

B8.5	Do landholders actively take part in cadastral surveying		(5=very actively, 1=not at all)
B.9	**Parcel and index maps**		
B9.1	How important are maps for land administration activities in your Woreda?		(5=very important, 1=not at all)
B9.2	Are all cadastral maps of your area connected to the national grid?		(5=fully connected, 1=not at all)
B9.3	Do you produce digital data sets and maps in your Woreda?		(5=fully digital, 1=fully manual)
B9.4	Is there demand for maps for uses other than land administration		(5=very high, 1=not at all)
B9.5	How involved are landholders in your Woreda in the production of cadastral maps?		(5=very actively, 1=not at all)

Please use this space to provide any comments on this topic that you feel may be helpful:

C	Land Value		
C1	**Valuation for compensation**		
C1.1	Do expropriation conducted only for projects planned for public purposes?		(5=always, 1=never)
C1.2	Is public benefit clearly defined?		(5=very clear, 1=not clear)
C1.3	Is the amount of money payable to evicted landholders is fair enough?		(5=fair, 1=very small)
C1.4	Is compensation always paid before eviction from their land?		(5=always before, 1=never before)
C1.5	Do communal holdings get fair compensation during expropriation?		(5=fair, 1=very small)
C1.6	Is the procedure for expropriation and compensation clear enough?		(5=very clear, 1=not clear)
C2	**Mass valuation**		
C2.1	How frequent is mass valuation conducted in your Woreda?		(5=every year, 1= not at all)
C2.2	How often are records of land administration used for mass valuation?		(5=always, 1= never)
C3	**Alternative livelihood creation**		
C3.1	How often development projects are prepared by your office to address the needs of evicted landholders?		(5=for all evicted, 1=never)
C3.2	Do affected people participate in project formulation?		(5=always, 1=never)
C3.3	Are evicted landholders happy after getting compensation?		(5=always, 1=never)

Please use this space to provide any comments on this topic that you feel may be helpful:

D	Land Use		
D1	**Strategic land use planning**		
D1.1	Is all land in your Woreda covered by a strategic land use plan?		(5 =fully covered, 1 = none)

D1.2	Did you take part during strategic land use preparation of your area?		(5 =at all stages, 1 = never)
D1.3	Is the strategic land use plan of your area legally enforceable?		(5 =fully enforced, 1 = never)
D1.4	How often is the information in strategic land use plan used as a guide for development in your Woreda?		(5 =always, 1 = never)
E	**Development control**		
E.1	Is the system for land use change control effective?		(5=effective, 1=not effective)
E.2	Is the system in place to enforce integrated watershed development plans		(5=sufficient, 1=no possibilities)
D7.3	What are three main weaknesses of land administration system?		
	Answer 1:	Answer	
	Answer 2:	Answer	
	Answer 3:	Answer	

Please use this space to provide any comments on this topic that you feel may be helpful:

Thank you for taking the time to complete this questionnaire

The researcher

10.6 Accuracy Comparison of AM Point Coordinates and PPP Measurements

Point id	E ast err	Nort err	sq err	sq err
ADET2650	0.026	0.039	0.001	0.001
ADET3030	0.013	0.005	0.000	0.000
ADET3040	0.007	0.018	0.000	0.000
ADIS0050	0.047	0.104	0.002	0.011
ADIS2610	0.022	-0.036	0.001	0.001
ADIS2640	0.016	-0.048	0.000	0.002
ADIS3051	0.007	0.098	0.000	0.010
ADIS3060	-0.124	0.201	0.015	0.040
ADIS3061	0.011	0.104	0.000	0.011
ALEM0850	-0.110	0.154	0.012	0.024
ALEM0860	-0.098	0.160	0.010	0.026
AYKE0410	0.076	-0.083	0.006	0.007
BUHR0420	0.094	-0.137	0.009	0.019
BUHR0430	0.082	-0.134	0.007	0.018
BURE0050	0.055	0.088	0.003	0.008
BURE3030	0.061	0.070	0.004	0.005
BURE3040	0.052	0.070	0.003	0.005
BUSO0860	-0.144	0.073	0.021	0.005
BUSO0880	-0.129	0.070	0.017	0.005
CHAC0850	-0.168	0.184	0.028	0.034
CHOK3050	-0.029	0.057	0.001	0.003
CHOK3060	-0.056	0.064	0.003	0.004
DABA0430	0.019	-0.125	0.000	0.016
DEBA0430	-0.011	-0.154	0.000	0.024

Point id	East err	North err	sq err	sq err
DEBA0890	-0.008	-0.148	0.000	0.022
DEBA2590	0.094	-0.007	0.009	0.000
DEBA2600	0.079	0.002	0.006	0.000
DEBA2640	0.091	-0.004	0.008	0.000
DEBA2650	-0.144	-0.212	0.021	0.045
DEBA3030	0.021	-0.007	0.000	0.000
DEBR0860	-0.031	0.104	0.001	0.011
DEBR0880	-0.088	-0.014	0.008	0.000
DEBR2610	-0.005	-0.026	0.000	0.001
DEBR2640	-0.042	-0.023	0.002	0.001
DEBR2641	-0.091	-0.024	0.008	0.001
DEBR2650	-0.026	-0.023	0.001	0.001
DEBR2651	-0.085	-0.021	0.007	0.000
DEBR3050	-0.044	0.067	0.002	0.005
DEBR3060	-0.126	0.086	0.016	0.007
DEJE0860	-0.026	0.142	0.001	0.020
DEJE3060	-0.078	0.117	0.006	0.014
DELG0410	0.079	-0.054	0.006	0.003
DELG0430	0.069	-0.054	0.005	0.003
DELG2600	0.072	-0.060	0.005	0.004
DELG2810	0.039	-0.093	0.001	0.009
ESTI2650	-0.028	0.001	0.001	0.000
GISH3030	0.037	0.040	0.001	0.002
GISH3040	0.015	0.031	0.000	0.001
GOND2600	0.052	-0.082	0.003	0.007
GORG2590	0.118	-0.007	0.014	0.000
GORG2591	0.061	-0.025	0.004	0.001
GORG2600	0.033	-0.062	0.001	0.004
HAMU2610	0.022	-0.011	0.000	0.000
HAMU2650	0.032	-0.005	0.001	0.000
HAYK0860	-0.156	0.049	0.024	0.002
HAYK0880	-0.180	0.034	0.032	0.001
IBNA0890	-0.010	-0.042	0.000	0.002
IBNA2610	-0.025	-0.039	0.001	0.002
IBNA2640	-0.007	-0.046	0.000	0.002
INFR2600	0.022	-0.069	0.000	0.005
INFR2610	0.077	-0.060	0.006	0.004
INJE3030	0.050	0.037	0.003	0.001
JIGA3040	0.008	0.080	0.000	0.006
JIGA3050	0.051	0.080	0.003	0.006
KEMI0850	-0.168	0.106	0.028	0.011
KEMI0860	-0.171	0.109	0.029	0.012
KOBO0880	-0.151	-0.039	0.023	0.002
KUMA0410	0.137	-0.077	0.019	0.006

Point id	East err	North err	sq err	sq err
KUMA0420	0.128	-0.080	0.016	0.006
LALI0880	-0.102	-0.029	0.010	0.001
METE0410	0.186	-0.101	0.034	0.010
METE0420	0.174	-0.103	0.030	0.011
MOTA2650	-0.008	0.048	0.000	0.002
MOTA3040	-0.038	0.030	0.001	0.001
MOTA3050	-0.035	0.030	0.001	0.001
SEKO0880	-0.097	-0.085	0.009	0.007
SEKO0890	-0.127	-0.082	0.016	0.007
SHAH0410	0.083	-0.038	0.007	0.001
SHAH2740	0.105	-0.032	0.011	0.001
SHAH2810	0.068	-0.025	0.005	0.001
SHEW0850	-0.197	0.167	0.039	0.028
SHIN0410	0.172	-0.065	0.029	0.004
TIKL0420	0.052	-0.109	0.003	0.012
TIKL0430	0.076	-0.097	0.006	0.009
WERH2600	0.025	-0.080	0.001	0.006
WERH2610	0.031	-0.077	0.001	0.006
WETE2740	0.057	0.031	0.003	0.001
WETE3030	0.051	0.013	0.003	0.000
WOGE0050	0.111	0.096	0.012	0.009
		Aver	0.008	0.007
		RMS	**0.088**	**0.084**

10.7 Root Mean Square Error for Different Observation times [in ± m]

Point name	1hr E	1hr N	2hr E	2hr N	3hr E	3hr N	4hr E	4hr N	5hr E	5hr N
ADD	0.083	0.049	0.082	0.056	0.083	0.053	0.083	0.053	0.083	0.051
ETJ	0.219	0.025	0.220	0.025	0.219	0.029	0.221	0.028	0.223	0.028
Aykel	0.005	0.026	0.005	0.027	0.005	0.025	0.006	0.022	0.006	0.021
Dabat	0.031	0.002	0.032	0.001	0.031	0.001	0.031	0.001	0.032	0.001
Delgi	0.001	0.022	0.001	0.017	0.001	0.017	0.001	0.018	0.001	0.017
Lalibela	0.001	0.038	0.001	0.034	0.001	0.040	0.001	0.040	0.001	0.041
Metema	0.008	0.128	0.009	0.112	0.008	0.124	0.009	0.131	0.008	0.132
Tkeldengay	0.016	0.018	0.016	0.017	0.016	0.015	0.015	0.013	0.017	0.014
Total	**0.36**	**0.30**	**0.36**	**0.288**	**0.36**	**0.30**	**0.366**	**0.30**	**0.37**	**0.305**

	3	9	5		3	5		6	1	
RMSE	*0.213*	*0.196*	*0.214*	*0.190*	*0.213*	*0.195*	*0.214*	*0.195*	*0.215*	*0.195*

10.8 Velocity Factor of AM Points

		Velocity geocentric (m/y)			Velocity local (m/y)		Speed per year
		dX	dY	dZ	E	N	(m/y)
Control	ADIS (IGS)	-0.0185	0.0184	0.0187	0.0260	0.0190	0.0322
Control	ETJI (CORS)	-0.0175	0.0191	0.019	0.0258	0.0192	0.0322
AM1-1	DEBANKA	-0.0188	0.0183	0.0188	0.0260	0.0192	0.0323
AM1-2	WETET ABAY	-0.0186	0.0185	0.0188	0.0260	0.0192	0.0323
AM1-3	KOSOBER	-0.0185	0.0185	0.0189	0.0260	0.0192	0.0323
AM1-4	BURE	-0.0185	0.0186	0.0189	0.0260	0.0192	0.0323
AM1-5	GISH ABAY	-0.0186	0.0185	0.0188	0.0260	0.0192	0.0323
AM1-6	JIGA	-0.0185	0.0185	0.0188	0.0260	0.0192	0.0323
AM1-7	ADIS ENA GULIT	-0.0185	0.0185	0.0188	0.0260	0.0191	0.0323
AM1-8	CHOKE	-0.0187	0.0184	0.0188	0.0260	0.0191	0.0323
AM1-9	DEJEN	-0.0187	0.0184	0.0187	0.0260	0.0191	0.0322
AM1-10	DEBRE WERK	-0.0188	0.0183	0.0187	0.0260	0.0191	0.0323
AM1-11	MOTA	-0.0189	0.0183	0.0187	0.0260	0.0191	0.0323
AM1-12	ADET	-0.0188	0.0184	0.0188	0.0260	0.0192	0.0323
AM1-13	HAMUSIT	-0.0190	0.0182	0.0187	0.0260	0.0191	0.0323
AM1-14	ESTIE	-0.0190	0.0182	0.0187	0.0260	0.0191	0.0323
AM1-15	DEBRE TABOR	-0.0191	0.0181	0.0187	0.0261	0.0191	0.0323
AM1-16	DEBRE ZEBIT	-0.0193	0.018	0.0186	0.0261	0.0190	0.0323
AM1-17	ADIS ZEMEN	-0.0191	0.0181	0.0187	0.0261	0.0191	0.0323
AM1-18	IBNAT	-0.0192	0.0181	0.0186	0.0261	0.0191	0.0323
AM1-19	INFRANZE	-0.0191	0.0182	0.0187	0.0261	0.0191	0.0323
AM1-20	WERHALA	-0.0192	0.0181	0.0187	0.0261	0.0191	0.0323
AM1-21	GONDAR ASTRO	-0.0191	0.0182	0.0187	0.0261	0.0192	0.0323
AM1-22	GORGORA	-0.0190	0.0183	0.0187	0.0260	0.0192	0.0323
AM1-23	DELGIE	-0.0189	0.0183	0.0187	0.0260	0.0192	0.0323
AM1-24	SHAHURA	-0.0187	0.0184	0.0188	0.0260	0.0192	0.0323
AM1-25	AYKEL	-0.0189	0.0183	0.0187	0.0260	0.0192	0.0323
AM1-26	KUMMER	-0.0188	0.0184	0.0188	0.0260	0.0193	0.0323

		Velocity geocentric (m/y)			Velocity local (m/y)		Speed per year
AM1-27	METEMA	-0.0188	0.0184	0.0188	0.0260	0.0193	0.0324
AM1-28	SHINFA	-0.0186	0.0185	0.0188	0.0260	0.0193	0.0324
AM1-29	TIKL DENGAY	-0.0192	0.0181	0.0187	0.0261	0.0192	0.0323
AM1-30	BUHRE (N GONDAR)	-0.0192	0.0181	0.0187	0.0260	0.0192	0.0323
AM1-31	DABAT	-0.0193	0.018	0.0186	0.0261	0.0191	0.0323
AM1-32	DEBARK	-0.0194	0.0179	0.0186	0.0261	0.0191	0.0323
AM1-33	CHACHA	-0.0189	0.0181	0.0186	0.0260	0.0189	0.0322
AM1-34	ALEM KETEMA	-0.0189	0.0182	0.0186	0.0260	0.0190	0.0322
AM1-35	SHOA ROBIT	-0.0192	0.0179	0.0185	0.0261	0.0188	0.0322
AM1-36	KEMISSE	-0.0194	0.0178	0.0185	0.0261	0.0188	0.0322
AM1-37	BUSO	-0.0192	0.018	0.0186	0.0261	0.0189	0.0322
AM1-38	HAYKE	-0.0195	0.0178	0.0185	0.0262	0.0188	0.0322
AM1-39	KOBO	-0.0197	0.0176	0.0184	0.0262	0.0189	0.0323
AM1-40	SOKOTA	-0.0197	0.0177	0.0185	0.0262	0.0189	0.0323
AM1-41	LALIBELA	-0.0195	0.0178	0.0185	0.0261	0.0189	0.0323
AM1-42	WOGEDAD	-0.0183	0.0186	0.0189	0.0259	0.0192	0.0323

Source: http://www.unavco.org/community_science/science-support/crustal_motion/dxdt/model.html

11 Table of abbreviations

AM-Amhara
ANRS Amhara National Regional State
ASCI-Amhara Saving and Credit Institute
AUPOS-PPP Online GPS Processing Service
BOFEPLAU Bureau of Environmental Protection Land Administration and Use
BOFED Bureau of Finance and Economic Development
CPR Common Pool Resources
CSAE Central Statistics Authority of Ethiopia
CSRSNRC Canadian Spatial Reference System, Natural Resources Canada
DEM Digital elevation model
DFID Department for International Development
DOP Dilution of precision
EEA/EEPRI Ethiopian Economic Association/ Ethiopian Economic Policy Research Institute
ELTAP Ethiopian Land Tenure and Administration Program
EMA Ethiopian Mapping Agency
EPLALUA Environmental Protection, Land administration and Land Use Authority (of ANRS)
EU European Union
FIG Federation Internationale des Geometres (International Federation of Surveyors)
FDRE Federal Democratic Republic of Ethiopia
FINIDA Finish International Development Agency
GCP Ground control point / Geodetic control point / Grid control point
GDP Gross Domestic Product
GIS Geographic Information System
GNSS Global Navigation Satellite System
GPS Global Positioning System
GTP Growth and Transformation Plan
HRS High Resolution Satellite Imagery
IGS International Ground Station
ISLA Information System for Land Administration
IT Information Technology
ITRF International Terrestrial Reference Frame
KA Kebele Administration (the lowest formal administration unit in Ethiopia)

LA Land administration

LAS Land Administration System

MF Micro-finance institutions

MoA Ministry of Agriculture

NASA-APPS National Aeronautics and Space Administration Automatic Precise Positioning Service

NGO Non-governmental organization

PBH Primary book of holding

PPP Precise Point Positioning

REILA Responsible and Innovative Land Administration

RENIX Receiver Independent Exchange Format

SBH Secondary book of holding

SDPRP Sustainable Development and Poverty Reduction Program

SNNP Southern Nations Nationality People

SIDA Swedish International Development Agency

USGS United States Geological Survey

UN United Nations

UN-ECE United Nations Economic Commission for Europe

FAO Food and Agricultural Organization

UNICEF United Nations Children's Fund

USAid- United states Aid agency

USD United States Dollar

12 Curriculum Vitae

Personal data

Name	Gebeyehu Belay SHIBESHI.
Date of Birth	September 23, 1968
Place of Birth	Gojjam, Finote selam
Country	Ethiopia
Nationality	Ethiopian
Marital status	Married
University	University of Natural Resources and Life Sciences

Education

Mar 2011- Sept 2014	Doctoral study at the University of Natural Resources and Life Sciences, Vienna, Austria
Sept 2005 to July 2007	Master of Science in Land management at the Royal Institute of Technology, Sweden
Sept 1999 to Aug 2002	Bachelor of Science in Agricultural extension at Alemaya University, Alemaya, Ethiopia
Sept 1984 to July 1985	Diploma in Forestry at Wondo Genet College of Forestry, Shashemene, Ethiopia

Professional experience

July 2009 to date	Land administration and registration expert
June 2007 to July 2009	Acting department head for land administration, surveying and registration department (BoEPLAU)
2002 to 2005	Team leader for land administration team (EPLAUA)
1998 to 1999	Surveyor
May 1987 to Sep 1989	Zonal watershed planning and community forestry expert (BoA)

Publications

Shibeshi, G., Fuchs, H. & Mansberger, R.(2014). Toolbox for the Development of Cadastral and Registration Proclamation for Second Level Certification Program in Ethiopia. International Journal of Sciences: Basic and Applied Research (IJSBAR) ISSN 2307-4531, 13(1), pp. 244-259

Shibeshi, G., Fuchs, H. & Mansberger, R. (2014). Lessons from Systematic Evaluation of Land Administration System of the Amhara National Regional State of Ethiopia: World Development, under review.

Shibeshi, G., Fuchs, H. & Mansberger, R. (2014). Formal and Informal Property Right Systems: The case of the ANRS of Ethiopia, Nordic Journal of Surveying and Real Estate Research, under review.

Shibeshi, G., Fuchs, H. & Mansberger, R. (2014). EVALUATION OF FITNESS OF THE NETWORK OF GROUND CONTROL FOR THE SECOND LEVEL CERTIFICATION PROGRAM IN ANRS OF ETHIOPIA, Survey review, under review.

Shibeshi, G., Fuchs, H. & Mansberger, R. (2013). Participatory and Pro-Poor Land Administration System of the Amhara National Regional State of Ethiopia: Lessons and Evaluation of the Current Status. Washington DC. The World Bank.

Shibeshi, G. (2011). Cadastral Template (Based on the PCGIAP-Cadastral Template 2003) Ethiopia, Country Report. Cadastral country reports based on a jointly developed PCGIAP/FIG template. Established under UN mandate by Resolution 4 of the 16th UNRCC-AP in Okinawa, Japan in July 2003.UN endorsement for cooperation with UN-ECE WPLA, UN-ECA CODI, and PCIDEA. http://www.cadastraltemplate.org/

Poster presentation

Shibeshi, G., Poster on Participatory and pro Poor Rural Land Administration System: The case of the Amhara National Regional State, Ethiopia. Presented at SID-UNIDO-ADA Workshop on Development of Human Settlements- Challenges for Sustainability" December 18-20, 2013 Vienna International Centre Wagramerstrasse 5 Vienna, Austria 1220

Printed by Books on Demand GmbH, Norderstedt / Germany